MW00463138

The Age *of* Melt

Hachette Book Group supports the right to free expression and the value of copyright. The purpose of copyright is to encourage writers and artists to produce the creative works that enrich our culture. The scanning, uploading, and distribution of this book without permission is a theft of the author's intellectual property. If you would like permission to use material from the book (other than for review purposes), please contact permissions@hbgusa.com. Thank you for your support of the author's rights.

Timber Press
Workman Publishing
Hachette Book Group, Inc.
1290 Avenue of the Americas
New York, New York 10104
timberpress.com

Timber Press is an imprint of Workman Publishing, a division of Hachette Book Group, Inc. The Timber Press name and logo are registered trademarks of Hachette Book Group, Inc.

Printed in the USA on responsibly sourced paper
Text design by Joel Ruffier
Jacket design by Vincent James

The publisher is not responsible for websites (or their content) that are not owned by the publisher.

The Hachette Speakers Bureau provides a wide range of authors for speaking events. To find out more, go to hachettespeakersbureau.com or email hachettespeakers@hbgusa.com.

ISBN 978-1-64326-151-5

A catalog record for this book is available from the Library of Congress.

The Age *of* Melt

WHAT GLACIERS, ICE MUMMIES, AND ANCIENT ARTIFACTS

TEACH US ABOUT CLIMATE, CULTURE, AND

A FUTURE WITHOUT ICE

Lisa Baril

Timber Press | Portland, Oregon

To John

Glaciers have shaped roughly a third of the land area of the planet. How could they not shape the ways we move and think, honed as we are, on sharp arêtes, domed cliffs, and the U-shaped valleys between, the floury rivers and string lakes held tight in steep canyons?

—Gretel Ehrlich, *The Future of Ice*

Contents

The Emergence of Ice-Patch Archaeology

In September 2021, I attended a scientific conference called Frozen Pasts. I got lucky. This particular conference (there had been four previous Frozen Pasts conferences) was held in Pray, Montana, less than an hour from where I live at the northern edge of Yellowstone National Park. Frozen Pasts gathers a rare breed of archaeologist–those who visit the most difficult of alpine environments in search of ice, or rather what the ice contains. They are known as ice-patch archaeologists.

At the time I worked remotely as a science writer for Utah State University. My job entailed writing reports for the National Park Service on the current condition of a wide range of subjects from birds to rare plants, but I was never asked to write about archaeology. In the National Park Service, cultural and natural disciplines are kept separate. This separation has never sat well with me. Aren't our human stories important to understanding the environment? Archaeologists would shout, "Yes!" I knew this to be true, which is how I found myself at a conference full of archaeologists.

At its core, archaeology is the study of our past through the material objects our ancestors left behind. When most of us think of archaeology, our minds drift toward images of Indiana Jones risking life and limb in his quest to rescue ancient artifacts from the hands of evildoers.

This romantic vision of archaeology has been reinforced through dozens of novels and movies: stories of lost cities, ancient temples hidden under a thick tangle of rainforest, and tombs of once-powerful rulers buried with priceless treasures. But archaeology isn't simply about objects, nor is it a treasure hunt; it's about using objects to interpret how people lived and interacted with each other and their environments in the past. Archaeologists are storytellers and artifacts are their muses.

Ice-patch archaeologists are especially attuned to the environmental aspect of not only our past, but also our present, because ice-patch archaeology arose out of our current climate crisis. As the climate warms, perennial patches of ice and snow (that is, bodies of ice that persist year after year) in mountain ranges around the world are melting. As these bodies of ice melt, artifacts that have lain frozen beneath the surface for hundreds or even thousands of years are beginning to emerge. But these aren't just the stones and bones that typically make up the archaeological record. Ice-patch artifacts include leather shoes and grass capes, birch bark containers, cups made of elm wood, and bows and arrows with bird-feather fletching still attached. Organic artifacts like these decay in a few years if left exposed to sunlight and wind, but alpine ice and snow have kept these objects in a state of suspended animation for hundreds or thousands of years. That is, until the ice began to melt.

The Disappearing Cryosphere

The frozen parts of the world are collectively known as the cryosphere. The cryosphere (from the Greek word *krios*, which means "cold") is made up of mountain glaciers, polar ice sheets, ice caps, sea ice, permafrost (frozen rock and soil), and seasonal ice and snow. Although only around 10 percent of Earth's surface is permanently frozen, the cryosphere stores almost 70 percent of all our fresh water. Nearly every drop (around 99 percent) is locked away in the ice sheets atop Greenland and Antarctica. The remaining 1 percent is held in mountain glaciers and perennial snow patches, and it is these bodies of ice that hold a surprising amount of human history.

Alpine glaciers and perennial snow patches are found all over the world, on every continent. Globally, the presence of glaciers and perennial ice patches varies by altitude and latitude. The farther north (or south) from the equator, the lower the permanent snow line. The foot of the Cotopaxi Glacier in Ecuador, for example, lies at a dizzying 16,400 feet, while some Alaskan glaciers flow directly into the sea. Aside from the global attributes of altitude and latitude, local topography within a given mountain range creates microclimates that influence where snow and ice can persist year-round. For example, in the northern hemisphere, north-facing slopes are colder than south-facing slopes. Even on warmer south-facing slopes, shadows provided by ridges, summits, and other landforms can help preserve snow and ice even during the hot summer months.

These perennial bodies of ice supply meltwater to river networks that extend far beyond the mountains themselves. Glacier-fed drainage basins cover a quarter of the land surface outside the poles, providing water to one third of the world's population, which is why mountain glaciers are said to be the water towers of the world. Glaciers are much more relevant to our everyday lives than we might realize. You could be part of the one third of the world's population who benefits from glacial meltwater, for example.

In the United States, glaciers supply water to people living in mountainous states like Montana, Colorado, and Wyoming, but also to those who live far from glaciers, such as in Texas, Missouri, Louisiana, and Oklahoma. Glacial meltwater irrigates fields and waters livestock in all these states. In Europe, South America, and high-mountain Asia, glacial meltwater generates hydroelectric power. Glacial meltwater keeps streams cool and provides habitat for fish and other wildlife, which cultures around the world rely on for sustenance. And nearly all mountain communities worldwide benefit from tourism to see glaciers, especially as they begin to disappear, sparking a new kind of adventure-seeker— the last-chance tourist.

Yet, worldwide, most of the more than two hundred thousand mountain glaciers are shrinking. Over the last two decades, mountain glaciers have lost almost as much volume as all the glaciers and ice sheets in Greenland and Antarctica combined, according to a 2021 study published in the journal *Nature*. And, according to the World Glacier Monitoring Service, for the thirty-fifth year in a row (since 1988), glaciers have lost ice rather than gained ice.

The future of ice is equally grim. A recent study published in the journal *Science* indicates that, based on the amount of greenhouse gases we've already put into the atmosphere, mountain glaciers will lose 26 percent of their mass—and half will disappear altogether—by 2100. We've already seen an average rise in temperature of 2.0°F (1.1°C). If we limit the rise in average global temperature to 2.7°F (1.5°C), which is considered an ambitious goal under the 26th Conference of the Parties (or COP26) agreement, then half of all mountain glaciers will still disappear. If we continue down our current path, we're headed for a 4.9°F (2.7°C) increase in temperature by 2100. At this level of warming, all the glaciers in the Alps, the North American Rockies, and New Zealand will melt sometime during the next fifty to seventy-five years.

It's hard to overstate what this loss means. Glaciers and permanent patches of snow are essential to the balance of life that so many cultures and communities—in fact, all humans—rely on to survive. That sounds rather dramatic, but the absence of ice means that the Earth is hot, perhaps too hot for survival. On July 4, 2023, as I wrote these pages, the Earth's average temperature reached its highest level since humans began collecting such data. So many temperature records have been broken in the last few decades and with such increasing frequency that the term "record-breaking" is almost a cliché. People all over the world are literally dying from the heat.

Although humans have adapted to live in multiple ecosystems and climates around the world, for the past six thousand years or so, humans have concentrated within a relatively narrow subset of temperatures that range between 52°F and 59°F (11°C and 15°C). A second, smaller mode occurs between 68°F and 77°F (20°C and 25°C), which corresponds to the region that experiences Indian monsoons (the rain helps mitigate some of the effects of high temperature). As a result of warmer temperatures, our habitable climate niche is moving higher in altitude and latitude. In the coming half century or so, an estimated one to three billion people will be left outside the human climate niche. Furthermore, many of the domestic plants and animals on which we have come to depend are adapted to thrive within the same climate niche, and most of the world relies on just a few major crops—corn, wheat, rice, and potatoes. We are hurtling toward a world that is largely uninhabitable.

Climate change is unequivocally human-caused. That is not up for debate and hasn't been for a long time, contrary to what some people would have you believe. We've been told a story that the climate isn't changing or that if it is, we are not to blame. We've even been told that warming is good. The narrative has been controlled by the few but powerful people who stand to gain from such a story. These powerful people have politicized not only the health of the planet on which we depend, but also our own health, by politicizing how water is shared and distributed as it becomes more limited, for one example.

But human-caused climate change couldn't be less of a political issue. It's a crisis for the global community of life—human or otherwise. The climate crisis, whether we live in an industrialized nation or one that has contributed very little to climate change, is part of our collective story. It's time we took the story back from the hands of those who would ruin our future so that they could profit in this one lifetime.

The Paradox of Melting Ice

Glaciers are our most visible manifestation of climate change and the effects of our actions on the one planet we have. We can see with our own eyes how fast the ice is melting. Photographs of glaciers from one year to the next reveal a vanishing feature of Earth's system—one that has been around throughout human history. But thinking about glaciers facing extinction reduces these bodies of ice to one story—melt. Yet glaciers have more than one story to tell. They figure prominently in Indigenous narratives, they inspire art and literature, they spark both fear and awe. They give and take life. Glaciers can be read like books, allowing us to peek behind the curtain of time. Scientists study ice to understand past climates, document major events in Earth's history, and chart our own influence on the planet.

Ice patches, too, are archives of the past. And now both archaeologists and climate scientists are learning that these seemingly insignificant bodies of ice have their own stories to tell. While objects melt out of glaciers all the time, they are often smashed to pieces because they are in perpetual motion owing to the immense pressure of their own weight and gravity. Perennial ice patches, on the other hand, never get large enough to begin flowing. Ice patches are stable, and it is their stability that makes ice patches so good at preserving ancient artifacts.

Ice patches have been overlooked as places with archaeological potential because no one thought that ice patches were that old. Not much had been found in or near them save for a few objects reported during the early twentieth century. Under the headlines "Ice Gives up Indian Arrow" and "Observable Finding of Centuries Old Weapon Found in North America," the March 15, 1925, issue of the Vancouver newspaper *The Province* reported the discovery of an intact arrow with fletching, sinew lashing, and a chipped stone projectile point on a glacier in North America. Around the same time, complete arrows with fletching, sinew lashing, and projectile points were found in the Oppdal

15

Mountains of central Norway. But these early discoveries were regarded as "mere curiosities and not the harbingers of a soon-to-be globally relevant research frontier," as one group of archaeologists put it.

It wasn't until a big melt in the early 1990s that archaeologists began to pay attention to alpine and high-elevation ice patches. It began with a mummy named Ötzi—a nearly perfectly preserved corpse that was discovered in 1991 in the Ötztal Alps, which straddle the border between Italy and Austria. The extraordinary circumstances in which this mummy was found and preserved led many archaeologists to believe that he was a one-off discovery, but at least one archaeologist suggested that the discovery signaled the start of something new.

In 1992, archaeologist Werner Meyer wrote that these finds "are hardly to be interpreted as unique exceptions, but rather as harbingers of further find assemblages, the discovery of which will be reserved for a future branch of science, the coming 'glacial archaeology.'" (Archaeologists in Europe and Canada refer to ice-patch archaeology as "glacial archaeology," even though their focus is on patches of stable ice rather than glaciers. Throughout this book, I use the term "ice-patch archaeology.")

Still, not everyone was convinced that ice-patch archaeology was truly a new branch of archaeology or whether it was simply traditional archaeology practiced in a different place—at the edge of alpine ice. But as the climate continued to warm and the ice continued to melt, artifacts began emerging from ice patches around the world, including in Switzerland, Canada, Norway, and most recently, Mongolia. The phenomenon became so widespread that by the end of the aughts, there was little doubt that ice-patch archaeology was unique.

Part of what sets ice-patch archaeology apart from traditional archaeology is its intimate connection to climate change. For organic artifacts to have been preserved in the first place, there must have been ice at the time the objects were left behind. And for those objects to have persisted, the ice must have been continuously or near-continuously present. But for an archaeologist to find such an object, the ice must have melted enough to expose it but not so much and for so long that the organic parts disintegrate. All of these factors converge to bring about a unique type of archaeology directly related to climate change.

Ice-patch archaeologists are responding to the climate crisis like emergency-room doctors—triaging ice patches by focusing on those that yield (or are most likely to yield) the most cultural material so that the loss of these heirlooms isn't total. It's an ambitious task given that there are only a handful of ice-patch archaeologists, hundreds of thousands of perennial ice patches worldwide, and a rapidly warming climate.

Ice-patch artifacts provide not only a portal to the past, but also a connection to the present that helps us think about our future—a future without ice. Within the ice is a cultural record of adaptation and response to past changes, but we're losing that record at a time when we need it most. It's a paradox: The more the ice melts, the more we learn about the past...while melting ice compromises our future. Although we can't rewind the clock to a time before human-caused climate change, we can use the knowledge gained from melting ice to help us respond more thoughtfully when considering the kind of future we want for ourselves and for the generations of humans yet to be born.

While it's tempting to think of ice-patch archaeology as a silver lining to climate change, its very existence also signals its own demise—because when the ice is gone, whatever organic objects were preserved within it will disappear forever. There will be no way to study ice because there will be no more ice to study. Hundreds, perhaps thousands, of ice-patch artifacts have already decayed from exposure after the ice around them melted. As one ice-patch archaeologist once told me: "It is a race with an uncertain finish line."

Stories in the Ice

In the following pages we will travel the globe, traversing both time and space, to discover what ice-patch archaeology has to teach us about our past, present, and future. We will travel to mountain ranges in places where ice-patch archaeology has been underway for thirty years or more—the Alps and Norway's Jotunheimen Mountains. We will also travel to the Canadian Rockies, the Beartooth and Absaroka Mountains in Montana and Wyoming, and Mongolia, where ice-patch archaeology has only recently emerged. In the Peruvian Andes, we will go on a spiritual journey to the Qolqepunku Glacier, and we will travel to the Himalayas, where locals are growing their own glaciers based on ancient traditions. Throughout these pages, I explore the shifting view that Europeans held of the Alps (from fear to awe to conquest) as they grappled with the effects of a changing climate on mountain glaciers during a period called the Little Ice Age. But the European understanding of the alpine region is only one way of relating to these rugged ice-filled landscapes.

In each of these places, we learn about the different stories ice has to tell, from hunting and subsistence to travel and trade to spirituality and reverence. While ice patches occur in similar environments around the world, the story they tell is unique to each region because of the extraordinary diversity of human culture. Some of these relationships were completely unknown before ice-patch archaeology arose, and they have challenged long-held assumptions about the role of mountains and alpine environments in culture.

As I listened to archaeologists, climate scientists, and Indigenous scholars speak at the Frozen Pasts conference, I began to see how ice-patch archaeology is especially well suited to shift the discourse on climate change. Ice-patch archaeology inherently draws people diverging from Western-conceived professions of scientific "expertise": people who are connected to the mountains, and to ice, because of their culture, because of storytelling, and because of ancestry.

I first learned about the phenomenon of ice-patch archaeology while working in Yellowstone National Park as an ornithologist. I never imagined I'd one day write a book about archaeology, but over the years I paid attention to what was happening with the ice and artifacts. Every week it seemed that I read another news story about another artifact melting out of the ice, whether it was an ancient arrow or a pair of skis. The climate-change connection combined with my love of mountains and interest in our human story inspired me to write this book. My hope is that readers begin to see glaciers and ice patches not just as melting bodies of ice, but as archives filled with stories and heirlooms.

ONE

Murder in the Alps

When the dead man arrived at the Institute of Forensic Medicine in Innsbruck, Austria, Hans Unterdorfer saw this was not a typical glacier corpse. Not only were all the dead man's limbs, ears, and toes intact and attached, but his body was also in a state of near-perfect preservation. Most corpses found melting out of glaciers have been ravaged by the shearing force of flowing ice—limbs wrenched from sockets, flesh torn from bones, skin stretched and misshapen. Glaciers sculpt landscapes, carve valleys, and crush boulders the size of small houses. A figure as fragile as a human stands little chance of emerging whole against such power. Yet here was this almost perfectly preserved mummy, dried out like a strip of jerky, weighing only about thirty pounds.

By the time the local undertaker delivered the mummy to the exam room, his flesh had begun to thaw. The stench of putrefying remains permeated the small space. Unterdorfer, the head surgeon at the university, heaved open the windows to diffuse the smell. He then rinsed away the grit and clumps of ice encrusting the dead man before beginning a brief exam. His secretary Uta took notes. At the top of the intake form, she filled in the date (23 September 1991) but left the name, occupation, sex, and address lines blank. She wrote:

> *Identity: not yet established; unknown mummified mountain corpse from the ... Ötztal [Alps]. The body, along with*

*grayish-black sodden material was delivered in a colourless
body bag and placed on the table ... The other items brought
with it are contained in a black bag.*

The "other" items included a small axe, several flint arrowheads, a
pouch attached to a belt—both made of leather—and a stone dagger.
The body had been found four days earlier by a German couple who
had been hiking on the south side of the Ötztal mountains in the Italian
province of South Tyrol. Erika and Helmut Simon were on their way
down the 11,529-foot Finailspitze—one among a row of needled peaks
that stitches northern Italy to southwestern Austria—when they saw
something odd breaking the surface of the snow. "Our first impression
was that it was rubbish, perhaps a doll, because by now there is plenty
of litter even in the high mountains," recalled Helmut. But as the couple
moved closer, Erika exclaimed, "Look, it's a person!"

It was a fortuitous accident the Simons were there at all. The day
before, they had hiked to the top of the Similaun—a massive mountain
southeast of the Finailspitze along the main alpine ridge. The trail
gains more than 6200 feet in just over five miles, rising far above timberline and alpine meadows to a place where only snow and ice cling
to bare rock and scree. To summit the peak, they had to cross a small
glacier that until the 1970s had been connected to the nearby Niederjoch Glacier. Just in the last ten years, the Niederjoch Glacier has lost
an impressive forty-six feet of ice from its leading edge.

During most summers, the south-facing glacier's gently sloping
tongue requires no technical skills and minimal mountaineering gear
to cross, but the summer of 1991 had been unusual. In early spring, a
plume of Saharan Desert dust swept across the Alps, staining the white
glaciers burnt orange. These Saharan dust storms are infrequent but
not uncommon in Europe (though they're becoming more intense as
the climate changes). A midspring snowstorm coated the dust in bright
white snow, but the shielding effect was temporary. By midsummer,
the spring snow had melted and unveiled the orange layer beneath.

The color of glaciers and other bodies of ice goes beyond aesthetics; it affects how fast ice melts.

Every surface exposed to sunlight absorbs some energy and reflects the rest. The amount of energy reflected by an object is called its albedo, which is measured as a percentage of total energy on a scale from 0 percent (all energy is absorbed) to 100 percent (all energy is reflected). Pure snow has an albedo of 80 or 90 percent, but old snow covered in a thin layer of dirt and debris might have an albedo of 40 percent or less. The lower the albedo, the faster the ice melts. The orange coating lowered the albedo of glaciers throughout the Alps, causing rapid melt.

The Simons paused at the edge of the ice to strap crampons to their boots. The crampons' metal teeth bit into the ice as the Simons stepped onto the glacier. Rivers of meltwater gurgled beneath their feet. Crevasses sliced deep, fading from glacial blue to black. Most crevasses were too wide to step over, so the Simons were forced to make long detours in search of crossings no wider than a stride length. It took so long to navigate the glacier that by the time the Simons reached Similaun's summit slender shadows drifted east as the mid-September sun slid behind the mountains.

As the couple took in the wide sweep of the Ötztal Range, Erika and Helmut considered their options. They hadn't planned on spending the night in the backcountry, but they knew they'd never make it back to their car before dark. The Similaun Hut, just a two-hour hike from the summit, offered food, plenty of beer and wine, and warm bunks protected from the cold alpine night. They figured they'd spend the night at the hut and hike out early the following day, but when they awoke, a slow-rising sun illuminated the Ötztal Range in the kind of vivid detail that only the driest alpine air can deliver. The day proved irresistible, so they headed up the Finailspitze rather than hiking down the mountain as they had planned.

They reached the summit around noon but didn't linger. Bruising winds rising from the valley below threatened to knock them off the narrow ledge. They pulled their layers tight around their necks to shield them from the chill and began searching for the marker that would lead them back to the trail and to the hut, where they had left a few belongings. Erika soon spotted a stick held up by a pile of rocks. These *steinmanndls*, or "little stone men," are sometimes placed in the alpine to prevent hikers from losing their way in landscapes where trails aren't always obvious. The steinmanndl was close to the Hauslabjoch—a pass not far from the main ridge between the Finail-spitze and the Similaun. They crossed a wide snowfield that ended along a natural rock wall as they made their way to the trail marker. The wall had collected a pool of meltwater from a nearby glacier. To avoid soaking their boots, the Simons were moving along the inside wall of the trench when Erika noticed something odd poking through the snow.

The corpse lay face down in the slush. Only the crown of his head and upper torso rose above the surface. His skin, tanned deep golden brown like old shoe leather, stretched thin across a pair of hollow shoulder blades and the knobby ridge of his spine. A few scattered objects lay nearby—a leather pouch, a small axe, and a scrap of birch bark in the shape of a tube that Erika described as "squashed flat, wound round with string or leather, and open at both ends." Erika thought a bird might have carried it there since they were well above the timberline.

The Simons were unnerved. Although they were keenly aware of the dangers of mountain travel, they'd rarely come face-to-face with the reality of what could happen in the mountains. Every year, hundreds of people die while touring the Alps. Some slip into crevasses. Others are buried in avalanches. A surprising number of people die of heart attacks. But the summer of 1991 had been noteworthy for the number of bodies recovered.

In the previous four weeks, five other bodies had been found melting out of Ötztal's glaciers. The first bodies belonged to a pair of novice mountain guides who died in 1953 after falling into a crevasse hidden by a weak snow bridge. A few weeks later, someone stumbled upon the corpse of a mountaineer who had died a decade earlier when the safety mechanism on his rope failed, releasing him into a 100-foot free fall down a crevasse. The oldest bodies found that summer belonged to a couple reported missing in 1934. Now a total of six corpses had melted out of the ice, the same number of bodies found in the Ötztal Alps during the previous thirty-nine years combined.

After the Simons informed the Similaun Hut's proprietor of their discovery, local police organized their fourth recovery of the season. Over the next four days, two dozen alpine recovery specialists, police, and curious locals worked at carving the dead man from the ice, but when he refused to budge, their tactics became more aggressive. They used a pneumatic chisel to chip away at the ice, chewing up the flesh around his hip. They used ski poles to gain leverage over the torso, letting the body slap back down onto the hard surface. They jabbed at the ice around him with a pick. They stepped onto his back, tugged at his limbs, and crushed his belongings under heavy boots. No one suspected that the body could be significant. Their main objective was to get the man off the mountain before autumn snow delayed the recovery until the following spring.

Their aggressive methods worked, and eventually, the corpse was ripped from the ice and loaded into a recovery bag with the zipper left partially open, because by now the body had begun to smell. The visible items lying on the snow were stuffed into a trash bag and hauled off the mountain for delivery to Unterdorfer's exam room.

* * *

On a cold metal slab at the Forensic Institute in Innsbruck, Unterdorfer bent over the mummy. The mummy's right lip curled toward his withered nose. His eyeballs were shriveled in their sockets like a pair of raisins. He was naked save for a leather shoe strapped to his right foot, which was stuffed with grass and held tight to a bony ankle with rope, string, or leather straps. Unterdorfer couldn't be sure. The deceased had also been wearing leather trousers, but those had fallen apart on the mountain. Scraps of fur-like patches clung to his body. Unterdorfer continued dictating his findings to Uta.

The corpse's right arm lay at his side, palm cupped as if he'd been "gripping a round object." The left arm "extended from the shoulder joint at an angle upward to the right," beneath his chin. A series of "blackish-grey linear discolorations ... arranged in parallel series" lined his right wrist and left elbow. The same markings inked his right knee, ankles, and lower back. The muscle tissue on his left hip had been "torn out in hide-like scraps," exposing his thighbone. Unterdorfer suspected scavengers, but the damage had been done during the difficult recovery.

Unterdorfer also noticed a patch of skin the size of a tennis ball that had been torn from the back of the corpse's skull, though he didn't see any fractures in the bone. Other than the exterior damage to the head, hip, and a few other places, Unterdorfer saw no immediate cause of death. From the man's unusual belongings, Unterdorfer suspected that the corpse was "many hundreds of years old ... an original inhabitant of the Ötztal," he joked. Judging by the corpse's presumed age, Unterdorfer knew that an autopsy was out of the question—at least for the time being. This was a case for an archaeologist.

Konrad Spindler, who was a professor of archaeology at the University of Innsbruck at the time, had been casually following the news about the dead man all weekend. On Saturday, the *Tiroler Tageszeitung*, a local newspaper, reported that a body had been found not far from the main ridge that marks the border between the Austrian province of Tyrol and the Italian province of South Tyrol. From an archaeologist's perspective, the six corpses that had melted out of the Ötztal Alps that summer were inconsequential. However, on Monday morning when Spindler read a second news report about the most recent body, it made him sit up. The morning news report read:

*The world-famous South Tyrolean mountaineers Reinhold
Messner and Hans Kammerlander, now on a circular tour of
the South Tyrol Peaks, took a closer look ... at the body found
at the Similaun Glacier. Messner remarked that the body
might be a warrior from the days of Frederick IV* [late fourteenth century], *popularly nicknamed "Empty Purse." Messner and Kammerlander ... believe that this is an exceptional
archaeological discovery.*

Spindler had recently established a subdepartment specializing in medieval and modern archaeology within the Institute of Archaeologies at
the university. A fourteenth-century corpse would fit in well with his
work, so when Rainer Henn, Unterdorfer's supervisor, called Spindler
about the dead man early Tuesday morning, he knew he had to see it
for himself.

Just after eight in the morning, Spindler met Henn at the Forensic
Institute. The brightly lit examination room, walled with pale green tiles,
reeked of disinfectant. The putrid odor of thawing flesh from the day
before had been mixed with the antiseptic smell of a research laboratory.
The corpse lay supine atop a metal slab. Some of the more unique objects
found with the body had been placed on a dark green cloth near the dead
man's head: an axe, a small flint dagger, a tassel of hide scraps attached
to a stone bead, a wooden stick with holes in it, a scrap of leather, and a
bag with just the tip of a flint point sticking out of it.

Spindler, tall with a receding hairline and round, wire-rimmed
glasses, bent over the table. He gave only a cursory look at the body. He
was more interested in the items brought down along with the body than
the body itself. (Archaeologists don't often work with human remains.
Their expertise lies in the objects associated with humans and what
those objects reveal about culture and life in the past.)

As Spindler handled each object, he considered its shape and form,
matching them in his mind to those of similar objects he'd seen in his
other work. Of course, he'd send samples of the man's body and his

belongings to a lab, perhaps several labs, for radiocarbon dating, but that could take weeks, maybe months.

It's worth mentioning here that a unique aspect of ice-patch archaeology compared with traditional archaeology is that so much of the material found is organic, which means that most ice-patch artifacts can be precisely dated. All organisms incorporate carbon isotopes into their tissues throughout their lives. Upon death, one carbon isotope (carbon 14) decays over time at a constant rate. By comparing the ratio of carbon 14 to stable carbon isotopes, scientists can determine the age of the organic material. This method works well for organic objects up to fifty thousand years old, after which the amount of carbon remaining in organic material becomes vanishingly small.

Although Spindler would wait weeks for the radiocarbon results to come in, he knew he could estimate when the man had lived by matching the craftsmanship and materials of his tools with similar tools of known dates. This is called "typological dating." Before radiocarbon dating of organic materials became available in the 1950s, it was one of the few means of dating that archaeologists had, and it's still commonly used by archaeologists.

The flint tools hinted at the Neolithic, but Spindler suspected that the axe dated to the early Bronze Age. He'd never seen a completely intact axe preserved with its binding and handle. No archaeologist had. Most often, the wood and bindings decompose quickly from exposure. Only a few axe heads had been found with their handles, but never attached to them. The axe handle, or haft, was two feet long, made of a single piece of wood cut from a yew tree and shaped like an L. The L was a branch that had grown out of the trunk and was cut while the tree was still alive. The blade of the axe head was less than four inches long and was covered in a thin ocher patina from prolonged exposure to water. It was wedged into a notch in the haft on the short end of the L and bound with leather cord. A scratch mark in the blade exposed a thin line of pinkish-orange metal. Spindler wasn't sure whether it was copper or bronze.

All was quiet in the exam room save for a clock ticking on the wall. Unterdorfer and Henn waited with anticipation. After some moments, Spindler spoke. The body is "about four thousand years old." He followed up more quietly with, "and if the dating is revised it will be even earlier."

In his 1996 book *The Man in the Ice*, Spindler recalls that Unterdorfer looked at him "with total disbelief." He had suspected that the corpse was many hundreds of years old, but four thousand years old was far beyond his wildest guess. Henn would later say that Spindler's "jaw simply dropped" when he saw the man's tools. This was a fully equipped prehistoric mummy, the likes of which had never been seen before. I imagine that the scientists erupted in a lively exchange of what this body could mean for archaeology in the Alps (and undoubtedly for their careers, especially Spindler's). No one in the room that day could have anticipated that the discovery would sprout an entirely new subfield of archaeology focused on ice.

For many years scientists believed that Ötzi had been found melting out of a glacier, but contrary to what scientists believed at the time, the body had not melted out of a glacier. If it had, his remains and belongings would have been shredded and his belongings turned to dust. It's possible he would not have been found at all. Instead, he'd been preserved in a stationary patch of ice in a mostly flat area next to but not part of a glacier. The fact that Ötzi had been preserved in a stationary patch of ice inspired the possibility that other ancient objects may also be frozen, intact, in the ice.

Later that day, Spindler and Henn held a press conference in one of the university's lecture halls. The press had been clamoring for interviews since reading Monday's report suggesting that the body was perhaps six hundred years old. Now that the date had been revised, reporters were greedy for an explanation of how a four-thousand-year-old, perfectly preserved corpse had lain so long near a popular trail without being discovered until now.

Although Spindler didn't have an immediate explanation for the mummy's existence, he tried to put the discovery in context by describing other ancient human remains. Spindler recounted the most famous mummies in Europe—the so-called "bog bodies." The moist, acidic soils of Northern Europe have preserved more than 1400 corpses from ancient times. Most of them date to the Iron Age—a time when the climate was favorable for bog development. The oxygen-free environment coupled with moist and acidic soils created a kind of "skin bag" with the remains of collapsed internal organs held inside. Scholars believe that many of these bodies were dumped into bogs after violent deaths for the biblical crimes of adultery or homosexuality.

Spindler also mentioned that in the late 1920s, the first of many burial sites were found in the Altai Mountains of Siberia. The bodies belonged to elite equestrian nomads of the Scythian Empire who dominated much of Eurasia from about the eighth to the third century before the Common Era (BCE). The bodies had been artificially mummified before burial, but their preservation was enhanced when water leaked into their tombs and froze. And in 1972, two hunters discovered eight freeze-dried mummies in two graves on the northwest coast of Greenland. The graves contained the bodies of six women and two children. The dry and bitterly cold arctic air kept these bodies from decomposing for five centuries.

When archaeologists find human remains, it's usually because someone buried them on purpose, and archaeologists knew where to look for them. While burials communicate a lot about the ritual aspects of a culture, those rituals may have little to do with how people lived their day-to-day lives. After all, death only happens once in a lifetime, and not everyone is treated the same after death. The Scythian mummies were buried with luxurious Persian carpets and Chinese silks, the remains of purebred horses, and richly decorated saddles. These items detail the riches of their culture—items that were probably enjoyed by a select few. But it's the ordinary that makes up most of our lives: the foods we eat, the clothes we wear, and the tools we use. It was the ordinariness of the Ötztal corpse that distinguished him from most other mummies and, ironically, made him that much more special.

As far as anyone could tell, the Ötztal corpse hadn't been buried, as one group of scientists would later claim. He'd apparently suffered an accident in the mountains and died surrounded by the tools he used every day. The Ötztal mummy was so well preserved by ice and snow that bits of brain clung to the inside of his skull. The undigested contents of his last meal were held within his shriveled stomach. And his tools, complete with wooden parts, emerged as whole objects. That an accidental burial could lead to such extraordinary preservation was unheard of at the time.

A few days after the press conference, Spindler and other university scientists established a committee to organize research on the mummy and his belongings, handle media inquiries, and ensure the continued preservation of the corpse. The committee purchased two freezers— one to store the body and the other as a failsafe in case the first freezer broke down. Both freezers were set to 21°F (-6°C) and high humidity to mimic the conditions of the ice on the mountain.

Next, the committee set to work on giving the dead man a name. Naming archaeological finds is a matter of practicality. Names are assigned based on the nearest geographic location included on official maps to where the object(s), or in this case a body, was found. The geographic eponym is then followed by the age or period the find is from, and finally, the type of find. The official name they gave the dead man was "Early Bronze Age glacier corpse from the Hauslabjoch." It was a mouthful and conveyed all the mystery of a dry scientific report, however useful to archaeologists.

Reporters craved a more exciting epithet, so they made up their own names for him. The French called him "Hibernatus." The Germans called him "Frozen Fritz." Others called him simply "The Iceman." But the nickname that stuck was "Ötzi"—a mash-up of Ötztal and Yeti—first printed in a Vienna newspaper soon after the discovery.

The moniker animated Ötzi for readers. This wasn't just an intriguing archaeological discovery; Ötzi had been a living person who had met an unfortunate end thousands of years ago. It could happen to anyone. It had happened to hundreds of people, including the five other souls found that summer. Although Ötzi had lived in a different time, his humanity was something everyone could identify with. Even the scientists who studied him began calling him Ötzi.

But it wasn't long before the committee had to revise Ötzi's official name. The forensic experts had sent bone and tissue samples as well as grass found with the corpse to four different European labs for radio-carbon dating. Those involved in the project also wanted to be sure that the dates of the items found with the body and the body itself matched. If so, then they could reasonably assume that the items belonged to Ötzi. Small samples of tissue and bone from Ötzi's torn hip were sent to labs in Oxford and Zurich. Labs in Uppsala and Paris received grass samples from Ötzi's clothes. While scientists in the Oxford lab were double checking their unexpected results, someone from the lab in Paris leaked the information to the press. Spindler found out about it from an Italian reporter who called him for comment.

Ötzi was 5300 years old—more than a thousand years older than Spindler had estimated. That put Ötzi into the Late Neolithic Age—the last of the stone ages, marking the transition from hunting and gathering to agriculture. In Europe, the Neolithic began roughly nine thousand years ago with the first farming societies in Greece. Farming spread throughout Europe, reaching the foothills of the Alps at least 6500 years ago. Given this new date, the mummy was officially renamed "Late Neolithic glacier corpse from the Hauslabjoch," but everyone still called him Ötzi.

* * *

In the months following Ötzi's discovery, the media grew increasingly desperate for his story. The committee had not given them much to work with. When the media didn't get the story that they craved from Spindler and his team, they began making up their own stories. Could he have been a prospector mining for precious metals? His axe, it was discovered, had been made of almost pure copper. The unusual material signaled that Ötzi was someone of status. One journalist speculated that Ötzi was a Neolithic businessman on his way to trade with a village on the other side of the main Alpine ridge. At one point, rumors swirled that Ötzi was a fake, planted on the mountain to create a sensation. Some went as far as to suspect the scientists of being in on the alleged conspiracy.

One particularly amusing theory was that Ötzi had tied one on and passed out on the mountain. In a letter to the editor published in the journal *Science*, George B. McManus wrote:

> *[Ötzi's] folded-over ear is a common consequence of having had too much to drink. Invariably, when one goes to bed in this state, he or she wakes up with a sore auricle [ear]. I thus propose that we cannot rule out the possibility that the Tyrolean man's downfall was indirectly attributable ... to overconsumption of prehistoric schnapps.*

Horst Seidler, a member of the Scientific Committee for Iceman Research at the time, responded in his own letter, also published in *Science*, that while a folded-over ear is a "consequence of alcohol abuse ... our colleagues from the Institute for Alpine History Research at the University of Innsbruck have ascertained that no prehistoric bottle of spirits was found among the ice man's provisions."

Part of the problem in studying Ötzi is that his body must be kept frozen, otherwise bacteria and fungi would destroy his remaining tissues. To prevent him from thawing too much, scientists working with Ötzi's remains were forced to fit their studies into brief thirty-minute windows. Those thirty minutes included unwrapping a set of sterile sheets from his body and re-wrapping them when they were finished. All this preparation meant that those studying Ötzi might only have a few minutes to collect samples, take X-rays, or conduct any other research they had planned. Given the time limits, it wasn't until nearly two years after Ötzi's discovery that the media heard what they so desperately craved, even if it was only partially true and impossible to verify.

In the summer of 1993, the University of Innsbruck hosted an international symposium on mummies. Scientists from all over the world came to speak about the human remains they studied, from coastal Peruvian mummies to the more famous Egyptian mummies, but it was Ötzi who drew the largest audience. After the lights dimmed in the auditorium, Spindler wove a captivating story about Ötzi's last days, filled with all the drama the media craved.

Spindler began with these words: "Our man had spent the summer before his death with herds of sheep and goats in the high pastures in the Upper Ötz Valley." Come autumn, suggested Spindler, Ötzi drove the flock into his village in the valley below. It was the harvest, and the village was busy threshing and storing grain, shearing sheep, and slaughtering some of the animals. While Ötzi busied himself with preparations for winter, he argued with someone from the village. It was a violent encounter. Perhaps there was a power struggle, or his lover took up with someone from the village while he was away tending sheep, or maybe he broke an important community rule. Spindler also theorized that Ötzi's village may have been attacked and that Ötzi had somehow escaped to the mountains, where he died of exposure in the shallow trench where he was found. Spindler described Ötzi's final moments like this:

> The trench offered some protection from the weather conditions, which were obviously an immediate threat to the man. [He] was in a condition of exhaustion. He knew that sleep meant death. To keep himself warm and awake he trudged a few paces up and down. His quiver fell to the ground, five yards away from the rock ledge. He must have staggered forward, completely exhausted, and frozen, he then tripped on a rock after

taking five steps and could not get back onto his feet. His cap fell off so that now he was bareheaded. With his last ounce of strength, he turned himself onto his left side, the least painful position for his injured rib cage. Then he fell into the sleep from which he was to awaken no more. Snow covered his body.

Spindler's story was unusually dramatic, even for an archaeologist whose job it is to tell stories about the past. He speculated about nearly everything, from the number of steps Ötzi took just before his death to his final thoughts. While many of the scientists in the room squirmed in their seats at Spindler's highly speculative story, the media were thrilled. Not only was Ötzi a shepherd from the Neolithic, but he was also a fugitive! Finally, they had a narrative they could work with. But while Spindler's tale was exciting, it wasn't Ötzi's real story. Ötzi's real story unraveled over the following three decades as radiologists, palynologists (scientists who study pollen), surgeons, and pathologists studied his body and belongings in more detail. Ötzi's true story was a far more brutal one, in fact.

Ötzi's Last Days

Ten years after Ötzi's discovery, the search for his cause of death finally moved beyond speculation. By this time, Ötzi had been moved to the South Tyrol Museum of Archaeology in Bolzano, Italy. A few weeks after Ötzi's discovery, surveys of the find spot revealed that the body had actually been found on the Italian side of the ridge and not the Austrian side as originally thought. This was a bitter blow to the Austrians. While the Italians began retrofitting an old bank in the heart of downtown Bolzano for Ötzi's eventual return, Ötzi remained in Austria.

Finally, in 1998, six years and four months after his discovery, Ötzi came to his final resting place at the South Tyrol Museum of

Archaeology, on the corner of Via Museo and Via Cassa di Risparmio. Ötzi resides on the museum's second floor along with his clothing and tools, but set apart from his belongings, in a room off the main exhibit hall. I visited Ötzi in July 2022. The darkened room had a single viewing window made of thick glass. Ötzi lay on the other side in a small climate-controlled chamber. A thin sheen of ice clung to his bony frame; museum conservators spray him with sterile water every two months to keep him hydrated. The glass table beneath him serves as a precision scale so that conservators can closely monitor changes in his weight that might signal too much or too little moisture. The temperature of his chamber is set to 21°F (-6°C).

I had seen so many photos of Ötzi that nothing about his appearance surprised me. He looked exactly as I expected. But to be honest, I was a little creeped out being this close to his remains. These weren't just bones. Ötzi was a real person. As I observed him, I couldn't help but wonder what he would have thought about all the fuss made over him. Would he consider his position on the second floor of a museum dedicated entirely to his life and death an honor or a curse? A crowd gathered behind me, so I only took a few more moments with him before exploring the rest of the museum.

In 2001, radiologist Paul Gostner and pathologist Eduard Egarter Vigl, both MDs at the General Hospital in Bolzano, scanned Ötzi's body using a mobile digital X-ray machine, once and for all unlocking the mystery of the Iceman's death. These were far from the first images collected of Ötzi. On six previous occasions, he'd been X-rayed and scanned using computed axial tomography, better known as a CAT scan. Images of his head, neck, abdomen, chest, spine, ribs, pelvis, and feet—pretty much his entire body—were taken by different research teams prior to the June 2001 scans. While these earlier images showed the damage done to Ötzi during the recovery effort, including a fractured left humerus and punctures to his pelvis and tibia from the pneumatic chisel, scientists also learned that Ötzi had suffered from a variety of painful conditions during his lifetime.

At some point, he'd fractured several ribs on his left side, which were well healed by the time he died. Ötzi had also suffered from degenerative arthritis in his spine, right hip, and knees. The "blackish-grey linear discolorations ... arranged in parallel series" that Unterdorfer noticed during his exam had been applied to all the same places affected by arthritis and were probably therapeutic. The tattoos, sixty-one in all, were drawn by rubbing pulverized charcoal into fine cuts to the skin. But while Ötzi had obviously led an eventful life, none of the injuries revealed in the scans and X-rays showed anything life-threatening—until the 2001 images.

When Vigl and Gostner studied the images, they noticed a strange triangular-shaped shadow in Ötzi's left shoulder. It was no bigger than the tip of a human thumb. To get a better look, they collected CAT scans of the area. Based on the images, the object had a density almost twice that of Ötzi's bone structure, so they knew it wasn't a piece of his skeleton. What's more is that the soft tissue surrounding the object was also denser than the tissue on the right shoulder, which suggested some kind of injury. And then they noticed a matching hole in Ötzi's shoulder blade and skin—Ötzi had been shot with a flint-tipped arrow. A high-resolution multidimensional CAT scan taken of Ötzi a few years later showed that the arrowhead had pierced Ötzi's subclavian artery. Essentially, he bled to death. Ötzi had been murdered.

Once they discovered that Ötzi had been murdered, they took another look at the images collected in previous years to see if they had missed anything. They had. Ötzi had a deep gash on his right hand consistent with a defensive wound. The CAT scan also showed dark spots at the back of Ötzi's cerebrum, which indicated that he'd suffered a blow to the head that rocked his brain against the back of the skull. But whether he'd been purposefully hit over the head or knocked it against a rock when he was shot with the arrow is impossible to know. There are no witnesses. It is the world's oldest cold case.

Not long after Ötzi's cause of death was finally determined, Klaus Oeggl, a paleobotanist and ethnobotanist at the University of Innsbruck, put together a possible chronology of Ötzi's last days by studying plant remains and pollen ingested with his last meals. Large amounts of pollen would suggest that Ötzi had intentionally ingested foods containing pollen while low amounts (but not so low that it could be due to random chance) of pollen indicate accidental ingestion. By looking at where in the intestines the various pollen types were found, Oeggl could estimate not only when but where Ötzi was when he ingested them.

To figure out the when part of the question, Oeggl turned to modern anatomy. It "takes 4 to 5 hours [for food] to pass from the gaster to the end of the small bowel (caecum), 9 to 12 hours to reach the transverse colon, and 14 to 55 hours to [reach] the end of the colon and rectum," wrote Oeggl. Since material that has passed from the stomach to the intestines contains not just food intentionally consumed but also background pollen, Oeggl was able to reconstruct the environment in which the meal was consumed.

The pollen grains in Ötzi's intestines suggested to Oeggl that Ötzi had eaten at least three meals in the days before his death. Based on the elevation ranges and climate preferences of trees and other plants, Oeggl surmised that Ötzi moved up and down the mountain during his last forty-eight hours. The following is how I imagine Ötzi's last days based on the last thirty years of studies.

On an early summer morning, Ötzi awoke to the rising sun. The night had been cold at nine thousand feet, but he was camped in a small alcove along a rock wall that protected him from the wind. A small fire kept him warm and discouraged predators from coming near. He added more fuel to the small fire and unrolled a moss wrapper that contained his breakfast—a mash of einkorn wheat, peas, and ibex meat. The day before he'd taken a serious fall while hunting and broken his bow in two. The fall tore his leather quiver and crushed all but two arrows. Without a bow, the two remaining arrows were useless.

He let the fire dwindle to a faint glow and began packing his things for the trip down the mountain to his village, where he planned to gather supplies. He pulled on a pair of fur leggings and a coat. Both were a patchwork of domestic and wild animals—goat, sheep, deer, and chamois (a species of mountain goat native to Europe). He tucked the tongues of his leggings into his soft leather shoes to keep them from filling with debris. He rolled up the grass mat that he slept on and stowed it in his wood-frame backpack. Across his shoulder he slung his quiver that held his two remaining arrows, and he pulled his bear-skin fur cap over his head.

Hours later he arrived at his village in the valley several thousand feet lower. At about forty-five, he might have been an Elder in his community. Few lived beyond such an age, but despite his many injuries, past and present, he could still tramp up and down the mountain with relative ease.

He gathered a long yew trunk that had been collected several years earlier from a young tree and left to season, as well as ten new arrow shafts. Some were from the wayfaring tree. Others were from cornelian cherry. Both made for good, strong weapons. He set to work on the arrow shafts, completing ten new ones. He still needed to make a flint arrowhead for each and affix feather fletching, but he moved on to the bow. Using a flint-tipped blade, he carefully trimmed the cut trunk of the flexible yet strong yew tree into a six-foot longbow. But before he finished, Ötzi was attacked. The encounter left him with a deep gash on his right hand when he instinctively raised it to protect his face. A dagger sliced him from thumb to forefinger, all the way to the bone. Blood poured down his forearm.

Ötzi managed to escape with his unfinished bow and arrows, axe, dagger, and a few other supplies. At the edge of the village, he washed the wound on his hand, packed it with birch fungus, and then wrapped it in bog moss. The moss would help staunch the bleeding and prevent infection. He pulled out a fistful of einkorn mash from his backpack and ate it quickly before climbing back up the mountain.

For six hours he climbed. It was near the solstice, and daylight lingered late into the evening. He arrived where he'd camped the night before and stayed the night. At dawn, he ate ibex for breakfast and then packed his belongings. He continued to climb. Perhaps he suspected he was being followed. He stopped just below a pass leading to the other side of the main ridge and rested near a patch of ice in a flattish area filled with snow.

While resting, he ate some bread and smoked meat. His hand throbbed and his knees ached. He never heard his enemy (or enemies) coming. Their leather shoes lined with grass were soft and quiet as they traversed the alpine terrain. The wind masked their approach. No marmots or squirrels called out in alarm, or else Ötzi didn't hear them. From this vantage, Ötzi's killer clearly viewed his quarry. He pulled an arrow from his quiver and drew his bow, aiming the point at Ötzi's resting form.

The flint-tipped arrow pierced Ötzi's left shoulder. Ötzi lurched to his feet and faced his enemy, gripping his dagger in his wounded right hand. But it was too late. His enemy was on him. Ötzi fell onto his back and knocked his head against the stone slab. The fall onto his back forced the arrow deeper into his left shoulder blade, piercing an artery. Blood poured from the wound as he gasped his last breaths. Within minutes he was dead. His enemy stood over him as Ötzi's eyes began to close.

Once the killer was sure Ötzi was dead, he rolled him over and pulled the arrow shaft from his back. Perhaps the shaft was engraved with the killer's signature mark, and he took it so that no one could identify the murderer. The arrow was never found. Everything else the murderer left as it was, including Ötzi's most prized possession—his copper axe. Possession of the axe would surely have given the murderer's identity away.

Ötzi, dead, lay on his stomach. His left arm was crossed awkwardly beneath him. His right hand still gripped his dagger. When his head hit the rock, his fur cap was flung from his head. It lay just in front of him, its chin strap broken. Within a few hours, Ötzi's body began to stiffen. Exposed to the arid alpine climate, Ötzi's skin dried in the summer sun and wind. Over time, most of the moisture left his body. His eyes sunk into their sockets and his brain shriveled inside his skull. His organs partially dried, and his skin shrink-wrapped around his bones, eventually sloughing off.

In all the months he lay exposed before winter snow concealed him, no animals scavenged him. If anyone found him, no one cared enough or had the will or means to bury him. Eventually, snow covered his body, entombing him in an icy grave. The ancient kingdoms of Egypt, Greece, and Rome rose and fell. Buddha, Jesus, and Mohammed lived and died. And still, Ötzi remained buried. The world changed in ways that Ötzi could never have imagined. And then one day, the summers grew warmer and the winters shorter. The protective blanket of snow and ice receded, and his body rose above the surface, exposed to the sun for the first time in thousands of years.

Ötzi's Legacy

Although scientists have learned much about Ötzi over the last thirty years, there are many details that are unknowable. We'll never know who killed Ötzi or why. We'll never know who missed him when he failed to come home. Nor will we know what kind of person he was—kind, jealous, violent? Although these are details that will remain unknown, new technologies may one day allow for more detailed understanding of who Ötzi was and how he lived. It is this possibility that keeps scientists returning again and again to his frozen corpse.

As I wrote the last sentences of this chapter, a group of scientists had just re-mapped Ötzi's genome. The first studies of Ötzi's DNA suggested that he most closely matched people living in Sardinia, but with more data and improved technology, scientists now know that Ötzi was related to early Neolithic farmers from Anatolia, in the region of contemporary Turkey, with a bit of hunter-gatherer genes mixed in. But beyond Ötzi's genetic origins, he also has something to teach us about legacy and inheritance.

There are people alive today who carry bits of Ötzi's DNA in their own, passed down from one generation to the next over many thousands of years. For all I know, given my European lineage, Ötzi could be one of my ancestors (although I doubt it). I wonder what he would have wanted for his descendants. He probably would have wanted a good harvest—for the snow to accumulate high in the mountains during winter and for that snow to melt slowly throughout the following summer to ensure healthy crops. He probably would have also wished for good health and prosperity for the next generation and the generation after that, and so on.

As a farmer, he would have been keenly aware of the relationship between how much snow fell in the mountains the previous winter and the size and health of the harvest the following autumn. He would have seen grain grow thick in wet (but not too wet) years and wither in years of drought. His life and livelihood depended on noticing changes in the environment and adapting to those changes.

Some of those adaptations included clearing forests for crops and livestock. Livestock changed the plant communities where they grazed; he and his contemporaries probably diverted streams to water agricultural fields. We've inherited the world he and his contemporaries helped shape. Although Neolithic farmers influenced how the landscape looked, their footprint was relatively minor compared to the one we leave today. If we could go back to Ötzi's time and see the world as he saw it, I think we'd mourn much more viscerally for the loss of what once was.

The discovery of Ötzi's remains not only inspired a new way of looking at the mountains; in fact, this one mummy, and all the ice-patch artifacts that would follow, signaled something else—the ice is melting. The glaciers that Ötzi was familiar with are actively disappearing, and it is not a natural change. And if we let glaciers disappear entirely, we will leave a far diminished world for those who come after us. Ötzi encourages us to ask ourselves: What kind of world do we want to leave for our descendants? What will our legacy be?

Before Spindler invented his Ötzi-as-shepherd story, he was often asked what the Iceman had been doing so high in the Alpine all those years ago. "Perhaps he wanted to visit his girlfriend in the next valley," quipped Spindler. Although he'd meant it as a joke, his reply called attention to a misconception in archaeologists' understanding of the relationship between alpine spaces and humans in the past—that people had avoided them. That would all change once the ice began to melt.

But just when did the ice begin its inexorable march toward oblivion? And how did mountains come to be perceived as inhospitable land-scapes to be feared? The answers can be found in Europe more than a century ago as culture, climate, and science collided in the foothills of the Alps during a period called the Little Ice Age.

TWO

How the Little Ice Age Shaped Western Views of Glaciers

In June 1644, Bishop Charles-Auguste de Sales traveled some sixty miles from Geneva to Chamonix, France, which lies in the shadow of the Mont Blanc massif. He came at the behest of Chamonix's tax collector, who visited de Sales a few days earlier bearing complaints from villagers that "their parish is situated in a mountainous valley, steep and narrow, at the foot of great glaciers which break off and descend ... causing such ravages that they are threatened with the entire destruction of their houses and property." Chamoniards reasoned that they shouldn't have to pay taxes on land destroyed by the "great and horrible glacier."

Chamonix's tax collector hadn't exaggerated the threat of Mont Blanc's glaciers. During the previous two years, the Les Bois tongue of the Mer de Glace—which has since disappeared—had advanced as much as 650 feet, plowing under farmlands and crushing homes. Before peaking in 1644, the Mer de Glace had destroyed more than 7500 acres of farmland throughout the parish of Chamonix, demolished twelve homes in Le Châtelard, and forced villagers in Les Bois to abandon the settlement altogether. Glacial refugees spilled over into nearby villages, adding further burden to the already poor farming communities living there.

Not only did advancing ice wreak havoc on farmlands and dwellings, but spring runoff from melting glaciers also caused disastrous floods and catastrophic outbursts of lakes normally held back by glacial moraines. As one resident put it, the ever-encroaching glaciers left villagers wondering "whether this [was] happening to them by divine permission as punishment for their sins." Just what sins the inhabitants of Chamonix thought they had committed to provoke Mont Blanc's glaciers is lost to history, but we do know that although these villagers weren't the first generation to witness the fury of advancing ice, they were one of the last.

When Bishop de Sales offered his prayers at the foot of the Mer de Glace and three other problem glaciers thundering down valleys, nearly damming the river Arve, the inhabitants of Chamonix and those across Northern Europe were living in the middle phase of what climate scientists call the Little Ice Age. Coined in 1939 by Dutch geologist F. E. Matthes, the phrase "Little Ice Age" was originally used to refer to an earlier four-thousand-year period of glacial expansion in the Sierra Nevada Mountains of California. But the term was soon repurposed to describe a shorter, 500-year cooling episode that began in the early fourteenth century and affected, to varying degrees, the entire northern hemisphere. During the Little Ice Age, temperatures dropped by 3.6°F (2.0°C). Summers were especially cool, which meant that glaciers didn't melt as much and retained more of the snow and ice that they accumulated during winter. As cooler summers prevailed, glaciers advanced for the first time in centuries.

Glaciers grow one layer at a time. As winter snow accumulates year upon year, it compresses the layers below in a granular mass of ice crystals. As additional layers of snow bury this granular mass, ice is formed, and it condenses. As the ice condenses, the spaces between the crystals become smaller. Eventually, the ice becomes so dense that a combination of pressure and gravity causes the ice to flow.

The difference between the amount of snow that accumulates during winter and the amount that melts during summer determines a glacier's mass balance. Mass balance is an indicator of a glacier's health. When more snow accumulates up high than melts down low, the glacier is said to have a positive mass balance, and it advances. When more snow melts than accumulates, the glacier is said to have a negative mass balance, and it retreats. While one or two years of negative or positive mass balance won't make a huge impact on a glacier's size or length, consistent mass balance measurements in one direction or the other will lead to major changes in a glacier's size, length, and volume.

While at times dreadful and deadly, the term the "Little Ice Age" was a bit of a misnomer. During a true ice age, great sheets of ice more than a mile thick spread across continents, plowing under rock and soil and pushing all life ahead of their snouts for hundreds of miles. Instead, the Little Ice Age was more of a climatic hiccup. Of course, the inhabitants of Chamonix weren't aware that they were living through a mini–ice age or that there had ever been such a thing as an ice age to begin with. All they knew was that the climate had shifted, and with that shift came hardship, especially in the already marginal farming communities of the Alpine foothills.

Prior to the Little Ice Age, generations of Europeans had enjoyed a 300-year-long period of comparatively warm, stable climate known as the Medieval Warm Period. Temperatures ranged between 1.3°F and 1.8°F (0.7°C and 1.0°C) above twentieth-century averages. Late spring frosts were virtually unknown to farmers. The warmer climate allowed them to grow crops and graze livestock in what had been marginal locations in mountain valleys like Chamonix. Vineyards flourished as much as three hundred miles farther north than they do today. So much wine came from England that vintners in France began to worry.

The warmer climate coupled with new technologies for plowing fields, sowing seeds, and harvesting crops led to a three-fold increase in crop yields. As surpluses increased, Europe's population more than doubled, tripling in some places. But the relative ease of the Medieval

Warm Period ended abruptly in 1315, when heavy and constant spring and summer rains made lakes of wheat fields. Come late summer, temperatures plummeted, and grapes withered on the vine. Livestock starved in their pens when hay failed to cure. Disease ravaged the survivors. As winter settled in, so did famine. That Christmas, farmers prayed for a better year, but heavy rains once again spoiled crops.

To understand what caused such an abrupt and dramatic shift in climate—and one that lasted centuries—I called paleoclimatologist Gifford Miller. Miller is a Distinguished Professor and is Director of the Center for Geochemical Analysis of the Global Environment (GAGE) at the Institute of Arctic and Alpine Research (INSTAAR) in Boulder, Colorado. He studies how interactions between the atmosphere, oceans, and ice influence climate. When I asked him about what triggered the Little Ice Age, he gave me an unexpected answer: volcanoes.

Along with ash, lava, and carbon dioxide, volcanoes launch sulfur dioxide into the atmosphere. While in the atmosphere, sulfur dioxide combines with water to form what are called sulfuric acid aerosols. These aerosols serve as tiny reflectors that scatter some of the incoming solar radiation back into space. If those sulfates are ejected with enough force to propel them into the stratosphere, about eleven miles above Earth's surface, then they are carried on wind currents around the world and the global climate will cool—at least until the aerosols settle out, which usually takes one to three years depending on the size of the eruption.

"We all know that explosive volcanism with a lot of sulfur will cool the planet for a short period, but there has to be something else in the energy balance to sustain colder temperatures," said Miller. That *something else* was sea ice. As warm tropical waters flow north along the coast of Europe, they meet cold Arctic water and lose heat. Prevailing westerly winds carry this lost heat across western Europe and are responsible for the relatively mild climate that is typical of the region. But during the latter half of the thirteenth century, the frequent replenishment of sulfuric acid aerosols from multiple volcanic eruptions cooled the climate just long enough to spur the growth of sea ice in the Arctic. This surplus sea ice had nowhere to go but south.

When this abundance of sea ice met with warm tropical water, it both cooled and freshened the ocean's surface, which slowed down the northward flow of ocean water. As sea ice cooled ocean currents along the North Atlantic, prevailing westerlies carried cold winds across Europe instead of the previously typical warm winds. This change in sea ice meant Europe's climate swung from relatively mild conditions to much colder conditions within a matter of years, plunging Europe into the Little Ice Age. And because sea ice has a high albedo, much of the incoming solar radiation was reflected into space, which led to further cooling. In other words, more sea ice meant colder temperatures and colder temperatures meant more sea ice. It became a self-sustaining feedback loop that was occasionally boosted by additional volcanic eruptions. There were ten major eruptions over the next several centuries. Miller told me that "once you had the sea ice start to expand and reflect more of the sun's energy, you ended up with this catastrophe for humans even though the forcing is really tiny."

Not everyone agrees with Miller on the details. Some climate scientists cite periods of low solar activity as a primary trigger. Across much of the world, three of the four coldest periods of the Little Ice Age roughly coincided with solar minima. But most scientists who've studied the Little Ice Age agree that cooling and freshening ocean currents in the North Atlantic played a major role in sustaining the Little Ice Age. While climate scientists debate the relative influence of one factor over another, the results are unequivocal. Cooler temperatures across continental Europe drove glacier growth in the Alps for the first time in centuries.

The combination of sporadic but major volcanic eruptions and the export of sea ice led to three major pulses in glacier growth during the Little Ice Age. The first pulse occurred in the mid-fourteenth century, the second in the mid-seventeenth century, and the last in the early- to mid-nineteenth century. While few firsthand accounts date to the first pulse, when locals sought tax relief during the second pulse, both ecclesiastical and civil authorities investigated their claims. They found that villagers were right. The glaciers were growing, and fast.

Chamonix's tax collector wrote that in 1616 the Argentière Glacier "every now and then goes bounding and thrashing or descending; for the last five or six years ... it has been impossible to get any crops from the places it has covered." In 1628, melting glaciers flooded the river Arve, rendering land adjacent to its banks useless for a time. And according to one report, a third of the arable land had been destroyed by avalanches and glaciers by 1640.

By 1644, the situation in Chamonix had become so dire that Bishop de Sales led three hundred Chamoniards to the tongues of four problem glaciers, blessing each one. The exorcism worked, but the spell was short-lived. Twenty years later, the Mer de Glace and other glaciers once again bounded down the valley. And as before, the Bishop of Geneva— this time Bishop Jean d'Arenthon d'Alex—was called upon to bless the glaciers in 1664, again in late summer 1669, and then again in 1690. But it hardly mattered. When the Mer de Glace finally retreated, one observer noted that the shrinking glaciers had "left the land they occupied so barren and burned that neither grass nor anything else has grown there."

The Devil had won.

Shifting Attitudes

The extreme conditions of the Little Ice Age inspired stories about mythical beings, monsters, and spirits who dwelled in the mountains. Witches were said to dance on the ice. They were blamed for unseasonable frosts and crop-destroying hailstorms. Many women were hanged or burned for the alleged crime of controlling the weather. Superstitions were especially prevalent in Switzerland—the most mountainous of all European countries—where tales of Frost Giants, whose offspring were avalanches, entertained children and adults alike. Alpine valleys were pressed into shape by the footsteps of Mountain Giants. Rivers were filled with the tears of their weeping wives and daughters. The dead were said to endure purgatory in glaciers.

Although many lowlanders viewed those who lived in the Alps as ignorant, superstitious rustics, educated city-dwellers were no less susceptible to superstition. In 1723, Swiss scholar Johann Jakob Scheuchzer, Fellow of the prestigious Royal Academy, published a guide to dragons of the Alps, writing that the best specimens "were to be found in the sparsely inhabited cantons of the Grisons [Switzerland]: that land so mountainous and well provided with caves that it would be odd not to find dragons there." He even speculated that wingless varieties were most likely females. During the Little Ice Age, glaciers themselves were sometimes depicted as dragons. In 1892, Henry George Willink drew the Mer de Glace as an open-mouthed dragon roaring down the mountain, devouring the land. Forked tongues of meltwater flowed from its open jaws.

But even as the Mer de Glace and glaciers throughout the Alps remained in their advanced states, and continued to advance with only minor retreats until reaching their maximum length in the mid-nineteenth century, outsiders grew curious about the Alps. Chamonix's first tourists were British aristocrats William Windham and Richard Pococke. They visited Chamonix in early summer 1741, well-armed and with a retinue of porters. This was unfamiliar country. It's hard to believe now, but people rarely visited the Alps back then, and those who lived there rarely left. The insulated Alps stoked trepidation in outsiders, but Windham discovered this apprehension to be unfounded. He later wrote in a letter to a friend that "the terrible description people had given us of the country was much exaggerated." Windham and Pococke found the locals to be obliging and welcoming.

When Windham and Pococke inquired about hiring guides to take them up the mountain, locals warned them that, owing to the difficulty, no one goes there except crystal seekers and chamois hunters, but they insisted they were up for the challenge. Besides, they had money, so a couple of local guides agreed to the task. Since the guides had little faith in the party's endurance for mountain travel, they brought loads of extra food, wine, candles, and fire starters in case they were forced to spend the night.

Early the following morning, they set out to climb Montenvers—the end of a long granite ridge overlooking the Mer de Glace, about three thousand feet above the valley floor. The route was so steep in places that they found themselves clinging clumsily to the rocks, but they made it to the top in half a day, much to the relief of the guides.

Windham was so enamored with the dramatic landscape that when he wrote about his trip he could hardly find words to describe what he saw. "We got to the top of the Mountain; from whence we had the Pleasure of beholding Objects of an extraordinary Nature. I am extremely at a Loss how to give a right Idea of it; as I know no one thing which I have ever seen that has the least Resemblance to it." Still, Windham did his best to describe the Mer de Glace. On the ice, "You must imagine your Lake put in Agitation by a strong Wind, and frozen all at once." The Mer de Glace translates to Sea of Ice.

When Windham published details of the trip two years later in two letters he wrote to his friend, he unwittingly ignited a tourism craze that continues to this day. Word quickly spread among Europe's elite, especially those in London. By the end of the eighteenth century, every wealthy European made a visit to Chamonix. It was considered part of an educational rite of passage called the Grand Tour. Soon, health spas and resorts sprang up at the bases of mountains throughout the Alps. They began as summertime retreats, but by the end of the nineteenth century, tourism included winter activities like ice skating, sledding, and the new amusement of skiing—a recent export from Norway, derisively called "plank-hopping" by cynics.

This newfound enthusiasm for mountains also drew geologists and naturalists to the Alps. Swiss naturalist Horace-Bénédict de Saussure summited Mont Blanc in 1787, after years of failed attempts. He followed the route taken a year earlier by crystal- and chamois-hunter Jacques Balmat and Dr. Michel-Gabriel Paccard. They had made the climb in response to Saussure's challenge, backed by a reward, to anyone who successfully gained the summit. These first documented climbs inspired the sport of mountaineering.

Saussure's interest in the Alps was largely geological. He learned from locals that the piles of rocks and debris found near glaciers were the "moraines of the glacier" writing that the "earth and rock types [were] driven forward by glaciers." He also observed that the glaciers were below the height of the moraines. Based on this observation, he theorized, correctly, that the glaciers had once been larger, and that a warmer climate was responsible for their reduced stature. His astute observations, however, were already common knowledge among locals.

Before the Alps became a tourist destination, local knowledge stayed within the bounds of the mountains. Although superstitions of Alpine spirits, giants, and witches were commonly held beliefs among locals during the Little Ice Age, those who lived among mountains were also observant of their environment. They made connections between glaciers, mountains, and climate that wouldn't be recognized by the outside world until formally educated scientists came to investigate for themselves, often relying on locals as guides while simultaneously dismissing their lived experiences. It was the peasants who truly discovered the ice age.

Discovering the Ice Age

Jean-Pierre Perraudin, born in 1767 in the Switzerland's Rhône valley, made his living as a farmer and mountain guide. From the top of the 10,919-foot Mont Fort, Perraudin was surrounded by snow-capped peaks, with views of iconic mountains like the Matterhorn and Mont Blanc. To him, everything he saw was God's creation, but he also saw creation of another kind: ice that moves mountains. His idea came from unusual boulders called glacial erratics.

Glacial erratics are rocks transported by the ice from one place to another. They could end up in a treeless field like the 18,200-ton Okotoks on the Alberta prairie. Or they could be a moss-covered boulder in a cedar forest on the Pacific Coast. Erratics are easy to pick out from other nearby rocks because they are made of different minerals than the native rock on which they land. They were literally plucked from the mountain and dragged away by the glacier. Erratics tell the story of the glacier's travels.

During Perraudin's time, though, there were many theories about the origin of erratics. The devout believed them to be deposited after the biblical flood. Others said that the Devil plucked them from the mountains and carried them to the plains with the intention of dropping them on unsuspecting villagers gathering for Sunday mass. Norwegians believed them to be evidence of troll fights.

Scientists also had their theories. Some imagined that glacial erratics simply dropped from the sky, although from where is hard to say. Or perhaps they had been ejected during massive volcanic eruptions. Another popular idea was that glacial erratics had been trapped in ice and carried to the Alps from the poles when much of the world was covered in a vast ocean.

In Ötzi's time, some glacial erratics became sites of worship. Many were engraved with abstract artwork, although some were adorned with recognizable symbols having to do with farming and hunting. Other erratics, like those at Stonehenge in the United Kingdom, were arranged in a circular pattern and capped with additional erratics dragged there from up to 150 miles away, which is a bit mind-boggling since some erratics at Stonehenge weigh more than forty tons. Although much about Stonehenge remains a mystery, archaeologists believe it was used as a burial ground and place of worship during the Neolithic and Bronze ages. Unlike those at Stonehenge, the erratics Perraudin noticed were all naturally brought down the mountain by the ice rather than intentionally moved (or engraved) by people.

Perraudin noticed that scars, or striations, on these large boulders in the foothills looked the same as those in the mountains. Not only did the scars match, but they were also made of the same rock type. Perraudin reasoned that the erratics had been transported from the mountains, but he knew that the great flood was not responsible. The only force he knew to be powerful enough to move boulders the size of houses was ice, and he reasoned that his backyard glaciers had been much larger in the past. But he was just a farmer, so no one paid his theories much attention.

In 1815, Swiss-German geologist Jean de Charpentier visited Val de Bagnes, a side canyon to the Rhône valley, to view the glaciers. He hired Perraudin as his guide. Charpentier spent the night in Perraudin's home and recorded part of their conversation in his journal. According to Charpentier, Perraudin told him:

> The glaciers of our mountains ... had in earlier times had a very much larger extension than today. Our whole valley, up to a very great height above the Drance [a river in Val de Bagnes], had been occupied by a vast glacier that stretched up to Martigny, as the blocks of rock [glacial erratics] that one finds in the environs of that town prove and which are too big for water to have been able to take there.

Although Charpentier agreed that water was an unlikely vehicle for transporting glacial erratics, he found it hard to believe that the glacier was once twenty-four miles longer than its present extent. Charpentier wrote of his conversation:

> Because [Perraudin] probably had never been beyond that town, and although I agreed with him on the impossibility of transporting erratic boulders by water, I nevertheless found

53

his hypothesis so extraordinary and even so extravagant that I
considered it as not worth examining or even considering.

Charpentier wasn't alone in his skepticism. Geologists were only just beginning to study glaciers; they knew almost nothing about them. Had scientists not been so skeptical of local knowledge, they might have advanced the field of glaciology much sooner.

Perraudin's observations might have gone nowhere had it not been for Ignaz Venetz, a more enlightened colleague of Charpentier's. Venetz was a highway and bridge engineer who frequently visited the Val de Bagnes. Conversations with locals helped him develop his theories on glaciers as agents of landscape change. He presented his ideas to the Swiss Natural Research Society in 1829. Venetz openly credited Perraudin and others for helping him formulate his own ideas regarding the past extent of glaciers and the transportation of erratics. Not surprisingly, Venetz's ideas were met with sharp criticism and dismissal. Charpentier even expressed embarrassment regarding his friend's attribution of local knowledge, writing that "it is known that in every country the inhabitants of the mountains are generally more inclined to superstitious beliefs than those in the plains." Eventually, Charpentier and other skeptics would come around, but not before several other prominent scientists, such as Louis Agassiz, expressed similar theories.

Louis Agassiz, a Swiss naturalist who began his scientific career studying fossilized fishes, took theories of glacial power a step further. At a geological conference in 1837, Agassiz proposed that great swaths of Earth were once encased in massive sheets of ice. He didn't use the phrase "ice age," but that is what he meant. He called glaciers "God's Great Plow." In his speech, Agassiz said that "a Siberian winter established itself for a while on a world once covered by lush vegetation and populated by large mammals. Death enveloped all of nature in a shroud, and the cold reached its extreme." It was a bold—and borderline heretical—statement at the time. Most scientists believed that the Earth had been gradually cooling down since its molten beginnings and that, in accordance with the Bible, it was no more than six thousand years old.

The age of the Earth came into question just a few years before Agassiz's grand speech, when Scottish geologist Charles Lyell published his three-volume work *Principles of Geology*. Lyell argued that Earth's layers were far older than anyone previously believed. If true, this meant that scientists no longer needed to compress human history into a mere six thousand years, and it opened the way for Louis Agassiz to publish his theories on past ice ages.

Unlike geologists and naturalists of the eighteenth and nineteenth centuries, who turned to mountains to investigate glacier movement, their contemporaries in archaeology rarely thought of mountains as worthy of study. One can hardly blame them. Archaeology was just finding its footing as a respectable branch of science. There were very few maps of the Alps, and those that did exist looked as if a child had scribbled some lines on a page. Most mountains had no name, and those that did were invariably named Accursed or Damned Mountain. And so, these first archaeologists were focused on more visible and immediately rewarding sites, like the ancient tombs of Egypt, the city of Pompeii buried under layers of ash, and ancient Greek ruins. The same year Agassiz gave his ice-age speech, Egyptologist Richard Howard-Vyse was busy blasting open the ancient pyramids of Giza. Besides, the Alps were covered in ice. What could possibly be up there anyway? But the ice-covered Alps were about to get a lot less icy.

Last Gasp of the Little Ice Age

On April 5, 1815, after several centuries of dormancy, Mount Tambora on the island of Sumbawa in present-day Indonesia (then the Dutch East Indies) belched out a thick, dark cloud, which hung menacingly over the top of the 14,100-foot peak. A thunderous roar echoed so loudly across the ocean that those living on the Molucca Islands, 870 miles northeast of Sumbawa, heard what they thought were cannons and gunfire. British colonists living on East Java, 1200 miles away, assumed they were under attack and began assembling their troops. Although the troops weren't familiar with the roar of a volcano, the island's Indigenous people probably were familiar with these sounds through either personal experience (the Indonesian archipelago is a volcanic hotspot) or their oral history, whether or not the current generation had experienced a volcanic eruption of this magnitude.

As the sun set on the evening of April 10, the whole mountain turned to fire as lava flowed down Tambora's flanks and out into the sea. British lieutenant Owen Phillips, arriving shortly after the eruption, recorded a firsthand account of that night. The raja of Sanggar told him:

> Three distinct columns of flame burst forth near the top of Tomboro Mountain ... and after ascending separately to a very great height, their tops united in the air in a troubled confused manner. In short time the whole Mountain next Saugur [Sanggar] appeared like a body of liquid fire extending itself in every direction.

By April 11, the worst of the eruptions had passed, but comparatively minor explosions continued through midsummer, with smoke emanating from the caldera until late summer. Ash and pumice stones rained down on the island, burying all of Sumbawa's plants, animals, and people in a layer of cinder at least two feet thick. In some places, the ash layer rose to nearly twice the height of an average person. Sumbawa and its neighboring islands lost more than ten thousand people from flowing lava and deadly gases. Survivors either starved or left, although life on nearby islands wasn't much better. An estimated one hundred thousand people living on Sumbawa and other islands died from the immediate effects of the eruption and food shortages in its aftermath.

The Mount Tambora eruption ejected nearly 110 tons of sulfate aerosols into the atmosphere and shaved nearly five thousand feet off the top of what had once been one of the tallest peaks in Indonesia. In its place, the explosions left a four-mile-wide caldera. It was the most powerful volcanic eruption of the last ten thousand years.

But what does lava have to do with ice? Remember that sulfuric acid aerosols act as climate coolers. As aerosols circulated throughout the stratosphere, the global climate cooled for the next few years. The eruption devastated crops around the world, leading to widespread famine and high food prices. In the northeastern United States, summer frosts led to mass migrations to the Midwest. In Europe, the summer of 1816 was the second coldest in the northern hemisphere (the coldest summer occurred in 1400) and was known throughout Europe as the year without a summer. It was also the year when writer Mary Shelley (then Wollstonecraft) visited the Alps from her summer retreat in Geneva.

Most of the summer had been so dreary, wet, and cold that she and her companions, poets Percy Bysshe Shelley (whom she married later that year) and Lord Byron, and Mary's stepsister Claire Clairmont, spent much of it indoors. When the skies occasionally cleared, Mary wrote that from the shores of Lake Geneva "rise the various ridges of black mountains, and towering far above, in the midst of its snowy Alps, the majestic Mont Blanc, highest and queen of all." Like most of her contemporaries, Mary desired a more intimate view of the mountains and their glaciers.

The Shelleys spent six days touring Chamonix. They traveled by mule with local guides who regaled them with stories of the Alps and the spirits and dragons that inhabited them. While neither believed the local legends, these haunting stories coupled with the sour summer weather inspired Mary's 1818 novel *Frankenstein*. In *Frankenstein*, Victor finally meets his creation on the Mer de Glace on a bitterly cold and dreary day. They retreat from the icy rain to a mountain hut on Montenvers where the monster (Shelley never gives him a name) tells Victor of his desperate loneliness over the two years since his genesis—a loneliness that drove him to murder Victor's young brother, William. Victor looks upon his creation with horror when he realizes what the monster has done.

Victor created a monster in a foolish attempt to control life and death, and by extension to control nature, but by the end of the novel he realizes that he never had control over his creation to begin with. Jane Nardin, former English and Literature Professor at the University of Wisconsin–Milwaukee, wrote of the novel "that it is Victor's own activities that have changed the natural world he inhabits from a place of beauty and goodness to one of looming evil and intractable pain." Victor's "desire to master the natural world ... and the impulse to adore its mystery and beauty are not fully compatible." Like Victor, we realize that it is our own activities that have changed (and continue to change) the natural world. We have introduced to the world a monster of our own creation, and now it seems we can't control the monster, because the monster is us.

Vanishing Dragons

Endings are often only appreciated in hindsight. When glaciers throughout the Alps began retreating in the latter half of the nineteenth century, no one suspected that it marked the end of an era. The Little Ice Age had not yet been named, much less described, and word of past ice ages had only just begun to spread, first within scientific circles and then more widely.

The retreat of glaciers coincides with the industrial revolution, which began in Britain during the mid-eighteenth century, spread to France in the early nineteenth century, and then spread to Germany and across the rest of Europe by the mid-nineteenth century. We traded doing things by hand to using coal-powered machinery, and the by-product of coal-powered machinery is black carbon.

The warming impact of black carbon is 460 to 1500 times stronger per unit mass than carbon dioxide, but because black carbon is heavy, it stays in the atmosphere for only one or two weeks before settling out. (Carbon dioxide can remain in the atmosphere for thousands of years.) When industrial black carbon (better known as soot) settled on glaciers, it reduced their albedo and accelerated melting. With the passing of air-quality laws in the United States and Canada in 1970, and in Europe a decade earlier, black-carbon emissions have decreased substantially in industrialized nations (although they have increased in countries with less-stringent clean-air regulations, such as China and India).

The industrial revolution makes the Little Ice Age the last notable and natural shift in climate before we assumed the role (unintentionally though catastrophically) as a major driver of climate change. And it is the benchmark against which all glacier retreat is measured. By the late 1890s, the Swiss had started tracking glacier fluctuations throughout the Alps. They hold some of the oldest records of glacier growth and retreat in the world. Key glaciers in the Alps, those with the longest records, lost more ice than they gained in every year since 2001. And since 1961, there have been only ten years in which glaciers gained ice. If, year after year, ice is lost rather than gained, eventually the Alps—including the majestic Mont Blanc—will be bare.

The Chamonix of today hardly resembles the poor parish it once was during the Little Ice Age. Downtown is packed with posh boutiques, five-star restaurants, and a dozen cheese-and-wine shops, patisseries, and chocolatiers. Climbers walk the streets still wearing their high-end mountaineering gear from a day summiting Mont Blanc. Those not inclined to climb the mountain on their own ascend in an overcrowded cable car, which whisks them to (almost) the top in mere minutes, leaving many dizzy from the sudden change in altitude.

I visited Chamonix during the second of three heat waves to sweep across Europe during the summer of 2022. On the day I visited the Mer de Glace, temperatures reached 99°F (37°C). During the first heat wave in June, the temperature on the summit of Mont Blanc, at nearly sixteen thousand feet, reached a record-breaking 50.7°F (10.4°C), shattering the previous record of 44.2°F (6.8°C) set in June 2019. The snow that had fallen atop the Mer de Glace the winter before had all but melted, exposing a layer of accumulated dust and dirt. In the eight weeks before my visit, the glacier had lost an astonishing ten feet of ice from its surface.

The heat in Chamonix was so intense that the thought of hiking three thousand vertical feet to the summit of Montenvers for a view of the Mer de Glace made me ill, so I rode to the top via the historic cog railway instead. From Montenvers, tour operators shuffled me to a cable car that whisked me up and over a forest of larch to a landing, where a set of metal stairs anchored to a granite sidewall led me down toward the glacier's terminus. If I had visited just fifty years earlier, I would have simply stepped off the lip of Montenvers and onto the Mer de Glace as William Windham and Richard Pococke did, but the glacier had retreated so much by 1988 that the tourism board installed the cable car and staircase to keep the glacier accessible. The glacier now lies almost five hundred vertical feet lower than it did just forty years ago.

Although the glacier was directly below me, it was hidden beneath a thick layer of debris. The first visible ice lay far up the valley in a crumpled torrent of rock and ice. Deep fissures and crevasses crisscrossed its surface. Although the Mer de Glace was only a fraction of its former size, I marveled at not only its enormity but also the size of the valley it created and all the rubble it has left behind. Glacial ice is so hard and dense that it is considered a type of rock. It is that density that not only carves mountains but is responsible for their ethereal blue and turquoise glow.

As I descended the staircase, I noticed deep striations that had cut the valley's otherwise smooth walls. Plaques commemorated the Mer de Glace's former glory. In 1988 I would have descended just three steps to reach the top of the ice. In 1990 I would have descended twelve steps. In 2000, 118 steps. In 2015, 370 steps. In 2022, after 580 steps, or about forty-two stories, I finally reached the glacier itself where an ice cave had been hollowed out of its terminus. Because the glacier is always flowing, the ice cave is built anew every year and has been for the last half century.

I stepped inside the glacier's throat and pressed my palm against its icy blue walls. Drops of meltwater fell from its palate. The ice-cold water was a balm against my skin. I thought about how long it took to create this glacier and the infinite number of snowflakes falling over millennia. Each one contributed to the layer upon layer of ice visible in the walls of the glacier. Its history was written and erased on the walls of the mountains as it advanced and retreated over the last ten thousand years. I was witness to a single moment in time of a landscape that took millennia to create.

Exploring the Alps began with the search for knowledge. Out of this search for knowledge grew a desire to conquer the world's tallest peaks. The Alps were an irritant—an obstruction to be overcome. While many people have climbed to the summits of Mont Blanc, Everest, and Denali (in Alaska), many have also died trying. But we could never quite tame those mountains, and so they stand even today as emblems of untamable wilderness. Even the Alps, with their gondolas and huts and hotels, aren't entirely tamed. The very steepness of their slopes; the extreme weather; deadly ice; and precarious footholds: none of these things can be wrestled into submission.

During the Little Ice Age, mountains were considered barriers to trade and travel. This perception colored how archaeologists would view mountains later on. What Agassiz and his contemporaries didn't know was that just as glaciers have ebbed and flowed over time, so have humans in response to the ice.

Rather than priests leading processions to halt the expansion of glaciers as they did during the Little Ice Age, they are now praying for their return. Reverend Fathers Charpentier and Thomas led Swiss parishioners to the edge of the Great Aletsch Glacier in 1653 to exorcise its evil spirits. In 1862, when the glacier advanced very near to the 1653 advance, villagers in Fieschertal and Fiesch began offering annual prayers to stop the glacier's progress. But by 2009, the glacier had retreated so much that locals worried it might one day disappear, so the local priest requested a change in the language of the official prayer from Pope Benedict XVI. "Glacier is ice, ice is water, water is life," is part of the new prayer, approved by the highest Catholic authority.

Climate-change deniers often use the Little Ice Age and the Medieval Warm Period to argue that Earth's climate has always been in flux and that the climate has been equally as warm in the past as it is today. This is true, except that the Little Ice Age and the Medieval Warm Period were specific to the northern hemisphere, whereas current climate change is a global phenomenon. And while it's also true that humans have dealt with climate changes in the past, none were as rapid or severe as current climate warming. The annual rate of increase in atmospheric carbon dioxide over the last sixty years is one hundred times faster than previous natural increases (like those that occurred at the end of the last ice age).

During the last eight hundred thousand years, carbon dioxide has fluctuated between 170 to 300 parts per million (or ppm for short). By 1950, we surpassed 300 ppm. In 2023, atmospheric carbon dioxide reached 420 ppm. By the time you read these words, the concentration of carbon dioxide will be higher still. The last time carbon dioxide was this high was three million years ago, during the Mid-Pliocene Warm Period. Temperatures were 3.6°F to 5.4°F (2.0°C to 3.0°C) higher than during the pre-industrial era. Sea levels were fifty to eighty feet higher. Humans (that is, members of the genus *Homo*) didn't yet exist. Humans emerged at the beginning of the last ice age. We are products of the last ice age. It's that deep-time relationship between ice and ourselves that we turn to now.

THREE

Our Ice-Age Origins

The Mer de Glace, like many alpine glaciers, is a remnant of the large ice caps that once crowned the world's tallest peaks during the last ice age. The last ice age, known as the Pleistocene, began 2.6 million years ago. During the Last Glacial Maximum, ice ruled the world. About twenty-seven thousand years ago, vast ice sheets more than a mile thick pressed down on North America and Northern Europe. In the southern hemisphere, all of Antarctica and large parts of South America and Asia were covered in ice. Ice spilled down New Zealand's Alps, the Andes, and the Himalaya Mountains. As much as 30 percent of Earth's land area was buried under a thick mantle of ice. With so much fresh water frozen into ice sheets and glaciers, the world was dry, windy, and impossibly bright. And yet, humans thrived.

During Earth's 4.5-billion-year history, there have been at least five major ice ages, beginning with the Huronian 2.5 billion years ago. The Huronian was followed by the Cryogenian, 720 to 635 million years ago. Then came the Early Paleozoic, between 460 million years ago to 420 million years ago. The Late Paleozoic followed the Early Paleozoic, occurring from 360 to 255 million years ago. And last but not least came the Pleistocene. Of the five ice ages, the Cryogenian and the Late Paleozoic are reputed to have been the most extreme, but it was during the Pleistocene that our genus emerged.

Our human family tree arose six to eight million years ago. While the oldest fossil belonging to our genus, *Homo,* is about two million years old (dating not long after the Pleistocene began), the oldest fossil belonging to our species is only about 315,000 years old. The fossil was discovered in Morocco in 2017, and it pushed back the emergence of anatomically modern humans—that is, a human that we'd recognize as kin—by more than one hundred thousand years. Some anthropologists think that our species arose at least half a million years ago.

Richard Potts, a paleoanthropologist and curator of the Hall of Human Origins at the Smithsonian National Museum of Natural History in Washington, D.C., says that throughout the six million years of hominin (the word for our human ancestors, including the Great Apes after the split from chimpanzees) evolution in East Africa, the climate gradually cooled. As the climate cooled it also became more variable. Potts observed that most of the major adaptations in our family tree occurred during periods of extreme climatic instability. These adaptations include adopting a completely bidpedal lifestyle, using roughly shaped stones as tools and then shaping stones for specific purposes, the use of fire to cook meat, and the development of larger brains. But why would these adaptations occur during periods of climatic instability?

Potts says that environmental instability and an unpredictable climate selected for behavioral and ecological flexibility in our ancestors. He calls it the "variability selection hypothesis." Central to this idea is that there was no single habitat, climate, or trend in climate that led to our most significant adaptations. Rather, it was variability itself that is responsible for our biggest changes. It works something like the following:

If the environment is stable with very little change over time, then a species only needs to be adapted to within a narrow range of variability. If the climate is changing toward a new stable state, a species must keep up with the change and adapt to a new range of variability to avoid extinction. But an environment that is constantly changing means that a species must adapt to survive a much broader range of conditions. Our ancestors had survived huge and rapid shifts in climate, and we acquired some of that adaptability, so that by the time modern humans arose, we were fully capable of dealing with pretty much any climate our planet had to offer, including the icy planet we ultimately inherited.

Prospering in the Pleistocene

Humans left Africa some sixty thousand years ago during a favorable climatic window, when the deserts of northeastern Africa and the Arabian Peninsula that we are familiar with today were wetter. A wetter climate provided green oases where hunter-gatherers could collect edible plants and harvest wild animals as they migrated north over many generations. Some of us went north into Europe. Others traveled east to Asia, eventually splitting in two with one group heading north and into the Americas and the other heading to the South Pacific islands, Australia, and New Zealand. Sometime between twenty-nine thousand and ten thousand years ago, humans had become a global species, and along the way our ancestors interbred with Neanderthals who were superbly adapted to cold climates. It's possible they passed on genes to help us survive the harsher climate. Our genetic makeup coupled with behavioral plasticity helped humans adapt to the myriad environments outside of Africa because our ancestors had been shaped by a constantly shifting environment.

Although human migration around the world was nearly complete (humans never colonized Antarctica), it wasn't directed. We had no goal to conquer the world. We were just really good at dispersing. We were also curious. Our ancestors must have wondered what lay around the next bend in a river, across the ocean, or over a ridge of granite peaks. Imagine setting sail in a dugout canoe not knowing what you'd find or if you'd survive. Many perished because of this curiosity, but that did not stop us. You might also include "adventurous" in our repertoire of traits that helped us disperse.

In part, ice helped us get around. During the height of the Pleistocene, sea levels were hundreds of feet lower than they are today. Lower sea levels exposed continental shorelines and created land bridges, including one route to the western hemisphere across the Beringia land bridge. Islands rose above the sea surface, enticing humans to sail from one to the other. Ice allowed us to expand our range to six of the seven continents. Even Antarctica has been claimed in recent decades by various nations for research purposes (although no one person lives there permanently).

Throughout our travels around the globe, we encountered novel ecosystems and animals we'd never seen before: mammoths, cave bears, wooly rhinoceros, aurochs, Irish elk, and steppe bison. All of them would go extinct by the end of the Pleistocene, but many continued to thrive. Humans grew to depend on species like reindeer, elk, red deer, horses, bears, and other mammals that survived not only a new predator (us) but a changing climate. We followed them as they migrated behind the vast ice sheets. We followed them as they rode the green wave of grasses and forbs up and into the highest mountains. We were curious, but we were also hungry.

The difficulty of living in an icy, dry climate—as humans who migrated out of Africa—forced us to develop new technologies, like using bone needles to stitch together animal hides. We wore them in layers to accommodate daily and seasonal changes in weather. We used animal hides to weatherproof rock shelters, building fires inside them that radiated heat off rock walls and kept us warm. We built tents with internal frames for easy transport. We shaped bone and stones to accomplish novel tasks that made life easier. We were, and still are, distinctly imaginative. Not only were our ancestors imaginative, but they were also cooperative.

Art, for instance, reveals a sophistication of behavior that can only be achieved by communities working together. In at least four hundred caves across Europe, our ancestors painted ice-age animals, leaving markings that some archaeologists believe signal the animals' breeding cycles. The information would have helped Pleistocene hunters plan hunts and communicate information to other hunters. Working together and communicating information helped us build social connections with distant groups of people. This network was essential to surviving the ice age, since people could seek help from their neighbors in times of hardship. Tens of thousands of years later, humans would begin to question (and answer) what caused the ice age.

The Greenhouse Effect

After Agassiz popularized his ice-age theory in the mid-nineteenth century, he and a few of his contemporaries speculated on the possibility of not just one but multiple ice-age cycles in Earth's past. Although most of us think of the last ice age as one long cold period, these periods ebb and flow like the tides. Ice ages are made up of many glacial and interglacial periods. The last glacial period of the Pleistocene ended 11,700 years ago. Although it may not seem so, we are still living in an ice age; it just happens to be an interglacial period within the Pleistocene known as the Holocene epoch. (Earth's newest proposed epoch, the Anthropocene, was rejected by the International Union of Geological Sciences.)

Some of the first evidence of multiple glacial cycles came from Scottish geologist Archibald Geikie. Geikie was knighted by Queen Victoria in 1891 for his vast body of work, which included finding fossilized plants sandwiched between layers of glacial till. The layers suggested multiple cold periods (indicated by glacial till) with warmer periods in between (when plants could grow, indicated by fossilized plants). In the following decades, the concept of four or five large glacial cycles of the last ice age emerged. Scientists now believe there were as many as twenty additional small glacial cycles during the Pleistocene.

Toward the end of the nineteenth century, Swedish chemist Svante Arrhenius was looking for an explanation for what caused these glacial cycles. How did the climate vary from icy cold periods to warm periods with less ice? His quest began as a side project to distract him from his crumbling marriage with former student Sofia Rudbeck. (Rudbeck was one of the first women to attend university in Sweden and was the first woman admitted to the Swedish Geological Society.) While Arrhenius worried over his failing marriage, he discovered one of the most important features of our atmosphere. By 1896, Arrhenius had observed that temperature fluctuated with atmospheric carbon—in other words, Arrhenius had discovered the greenhouse effect.

Carbon dioxide is a greenhouse gas that increases temperature by absorption and reflection of heat. Without it, Earth's average temperature would be below freezing. Too much carbon dioxide in the atmosphere and the Earth heats up, just as we're seeing today. Through laborious mathematical equations, Arrhenius worked out that by cutting atmospheric carbon in half, the temperature in Europe would decrease 7°F to 9°F (4°C to 5°C)—a reduction that would result in an ice age. But no one knew whether carbon dioxide could change that much, so he turned to his colleague Arvid Högbom.

Arvid Högbom, also from Sweden, studied carbon emissions from volcanoes and the ability of the oceans to absorb carbon dioxide (the oceans are the world's largest carbon sink). He then began to wonder about carbon emissions from all the new factories that had been built over the last century. Högbom figured out that human activities during the industrial revolution were adding carbon dioxide to the atmosphere at a rate roughly equal to the carbon dioxide added through natural processes like volcanic eruptions and wildfires. Armed with this knowledge, Arrhenius wondered what would happen to temperature if the amount of carbon dioxide in the atmosphere doubled through human activities. The numbers were startling. Doubling the concentration of atmospheric carbon dioxide would raise Earth's temperature by an average of 9°F to 11°F (5°C to 6°C). But based on the concentration of atmospheric carbon dioxide in 1896, which was about 295 ppm, a doubling of carbon dioxide would take about three thousand years.

Neither Arrhenius nor Högbom thought that a doubling of carbon dioxide was a problem. In fact, they believed it to be a good thing. Arrhenius postulated that a warmer planet was necessary to feed a growing population. Besides, they reasoned, three thousand years is a long time. Their main concern was not that the world would grow too hot but that it would grow too *cold*.

In 1908, Arrhenius published a popular science book titled *Worlds in the Making*, in which he allayed fears of "a new ice period that will drive us from our temperate countries into the hotter climates of Africa" by writing that "there does not appear to be much ground for such an apprehension" because "the enormous combustion of coal by our industrial establishments sufficed to increase the percentage of carbon dioxide in the air to a perceptible degree." If he only knew what was to come. The world was indeed already growing hot.

By the time Arrhenius wrote those words, the amount of carbon dioxide emitted through industrial processes had increased enormously and, as a result, Arrhenius was forced to revise his timeline. Three thousand years became a few centuries. But Arrhenius only mentioned this revised timeframe in passing. He still wasn't concerned about a warming climate. The potential for serious global warming seemed irrelevant.

Milankovitch Cycles

The amount of carbon dioxide in the atmosphere changes in response to shifts in the distribution of solar radiation on Earth. Scientists knew that the tilt of Earth's axis and the shape of Earth's orbit changed the distribution of solar radiation on Earth and that these changes affect climate, but no one had tied these changes to the ice age until the early twentieth century. Enter Milutin Milanković.

Milanković was born in 1879 to a well-heeled family in Yugoslavia (now Serbia). After earning a doctorate in 1904 for theoretical research on the design and material of concrete, he took a professorship at the University of Belgrade, where he taught mathematics for the next forty-six years. Milanković had a passion for communicating science to the public. During the 1920s, he wrote "letters" to an imaginary friend for a Serbian magazine, in which he visited earlier scientists and explored their ideas for a general audience. Many of his articles were about astronomy. Although a gifted science communicator, Milanković was an even more gifted mathematician.

Beginning in 1911, Milanković undertook the daunting task of mapping Earth's orbital cycles and how those cycles interacted with and changed the distribution of solar radiation on Earth over the last six hundred thousand years. It took him thirty years of painstaking, hand-written calculations to complete the task.

Earth's orbit changes in three distinct cycles of different lengths. The longest cycle is orbital eccentricity, which is the shift in Earth's orbit around the sun from a near circle to an ellipse. A full cycle takes one hundred thousand years to complete. The second cycle is obliquity, or the tilt of the Earth on its axis. Earth tilts between 22 degrees and 24.5 degrees on a 41,000-year cycle. Earth was at its maximum tilt at the end of the Pleistocene. The last cycle is called precession. Precession is the direction in which Earth's axis is pointed. The Earth wobbles on its axis over a 26,000-year cycle. These three cycles—eccentricity, obliquity, and precession—are now called Milankovitch cycles, and they overlap in ways that guide the onset, duration, and end of ice ages. But Milanković's work still couldn't be tied to the geological layers that Geikie identified, because there wasn't a good way to date the different layers of rock showing the stacked glacial and interglacial periods. Geologists needed to figure out when each cycle occurred and over what timeframe to see if they matched the calculations Milanković had predicted.

It wasn't until the 1970s that science had advanced enough to tie Milankovitch cycles to ice-age cycles. Scientists had developed methods for figuring out past climates from the fossils of microscopic zooplankton called foraminifera (or forams) living in the oceans. Forams build shells of calcium carbonate pulled from their environment. When forams die, their tiny bodies rain down to the ocean floor where they, along with the carbon dioxide in their shells, accumulate in layers.

The forams' calcium carbonate shells contain oxygen, which comes in two forms: light and heavy. The ratio of the light and heavy forms is an indicator of ocean temperature and correlates to how much ice existed when the shells were formed. When oceans are colder, shells contain more heavy oxygen, and when oceans are warmer, the lighter form dominates. The record left behind by these shells is extremely precise given the short life span (weeks to several years) of forams and the narrow climate niche in which different species of forams live. Ocean core sediments are the longest archive of past climate, extending sixty-six million years into the past. (Layers older than this have been destroyed by activities of plate tectonics, namely seafloor spreading and subduction.)

For the first time, Milankovitch cycles were explicitly tied to temperature and ice-age cycles. Unfortunately, Milanković died before his theory was validated. While Milankovitch cycles set the pace of ice-age cycles, the climate had been gradually cooling since long before the last ice age began. Curiously, the rise of the Himalaya Mountains may have helped trigger the Pleistocene.

A Possible Ice-Age Trigger

The Himalayas are the youngest mountain range on Earth, and they are still rising. Three hundred million years ago, India was part of the Gondwana supercontinent near the South Pole. As the large continent broke up, India slowly drifted north and eventually crashed into Asia. As India muscled its way into Asia about fifty million years ago, the crumpled and pushed-up rocks caught in the collision formed the Himalayas. The newly formed mountains shifted patterns of rainfall, gathering them into annual monsoon rains. As rain fell on the Himalayas, it wore down the rocks and reacted with latent carbon dioxide. Dissolved carbon dioxide and rainwater washed into the oceans, and they became fodder for making foram shells. This chemical weathering of mountains helped cool the planet by reducing the amount of carbon dioxide that would otherwise become available in the atmosphere.

As the amount of ice at the poles increased, albedo feedback led to cooler conditions, which led to more ice. In addition, cold ocean water holds more carbon dioxide than warm ocean water, so as the ocean cooled from the ice, it absorbed more carbon dioxide. And as the ice sheets grew, sea levels lowered, exposing more land area for carbon-capturing plants to grow, which further cooled the planet. Expanding sea ice also reduced upwelling that normally brings carbon dioxide to the surface. So, ice ages aren't simply the result of cooler temperatures; there are self-reinforcing feedback loops or cycles that help to not only spur the ice age but to sustain it—just like the export of sea ice helped sustain the Little Ice Age. These self-reinforcing cycles may also help explain the shift in glacial cycles, which lasted about 41,000 years during the first half of the Pleistocene but lasted around 120,000 years during the last half of the Pleistocene. It goes something like this:

During the early phase of the Pleistocene, continental ice sheets were relatively small. The ice age had only just kicked off, so not much ice had formed. Because the ice sheets were small, they may not have survived interglacial periods. They simply melted away. But as the Pleistocene progressed and temperatures continued to plummet, the ice sheets grew larger, which made them more resistant to melting during interglacial periods. The more ice there was, the greater the surface area for reflecting solar radiation and the higher the albedo, or reflectivity. Bright white surfaces like ice reflect far more energy than, say, a dark green forest. The high albedo across a large portion of Earth's surface helped to further cool the planet, which stimulated the growth of more ice. Again, just like during the Little Ice Age, glacial cycles during the Pleistocene were sustained by self-reinforcing cycles or feedback loops.

Another theory that may explain why glacial cycles lengthened during the latter half of the Pleistocene has to do with erosion. The physical force of glaciers grinds rocks into dust. As the ice sheets expanded, fine particles of dust became airborne. When that dust settled on glaciers, it reduced their albedo and therefore increased the melt rate. Eventually, the melted ice and dust drained into the ocean. This dust that was deposited in the oceans provided critical nutrients to ocean-dwelling animals, which stimulated the growth of algae, thus drawing carbon dioxide out of the atmosphere and cooling the climate. Some scientists think that the shift from 41,000-year to 120,000-year cycles occurred when glaciers effectively eroded most of the soil down to bedrock. With so little dust available to gather on the ice sheets, Earth got a whole lot brighter; there just wasn't much dust to settle on the glaciers and warm them up.

The shift in the timing of glacial cycles is probably a combination of these and perhaps other factors, but it's difficult to tease these variables apart, because as the ice sheets grew in size, they erased earlier evidence of smaller glacial advances. Less is known about the four earlier ice ages because the further back in Earth's history one goes, the more illegible the book of Earth becomes. But while the ice sheets erase their own history when they plow over rock and soil, the ice itself is an archive of the past.

Library of Ice

As layers of glacial ice accumulate over time, tiny bubbles of air become trapped inside. These trapped air bubbles are time capsules of the atmosphere from around the time the snow fell. By extracting the gases trapped in the ice core's air bubbles, scientists can determine what the climate was like in the past. In addition, sulfuric acid aerosols from volcanic eruptions, black carbon from wildfires and industry, and human-caused pollution from lead and the burning of fossil fuels all drift on air currents and settle on ice, adding to the wealth of information glaciers provide—at least as long as the ice doesn't disappear.

Today there is considerably less land covered in ice (around 10 percent) than during the Last Glacial Maximum. After the Pleistocene ended, mountains became the last refuge of ice outside the poles. Of all the different components of the cryosphere, the two remaining ice sheets—one that presses down on Greenland and the other that spills across Antarctica—hold 99 percent of all frozen fresh water. That's almost all of it. Imagine the United States and Mexico buried under more than a mile of ice, and you get a sense of just how enormous the Antarctic ice sheet is. Though smaller, Greenland's ice sheet is also massive at about three times the size of Texas. Together, these two ice sheets would contribute two hundred feet of sea level rise if they were to melt all at once. A two-hundred-foot rise in sea-level would alter the contours of the world's major land masses and drown island nations like most of those in the South Pacific. That would mean no more Paris. Venice is already almost underwater so it, too, would disappear. The Netherlands and Denmark—gone. Florida and the Gulf Coast—gone. San Francisco would become a collection of islands. All of Bangladesh and coastal India would be submerged. The Amazon River would become an inlet of the Atlantic Ocean. Plato's mythical Atlantis would become reality.

But it is not only the polar ice sheets we should be worried about losing. Alpine glaciers and ice fields atop mountains are also retreating. As the equilibrium line—the imaginary line on a glacier where accumulation equals melting—moves higher in altitude, the ice caps will begin to melt and the precious information these ice caps contain will disappear.

Although alpine glaciers like the Mer de Glace have been around since the Last Glacial Maximum, the ice they contain isn't from the last ice age. This is because ice that accumulates high up the mountain at the glacier's source flows down the mountain, where it melts. Glacial ice is constantly renewed. This constant renewal means that the ice is usually only about five hundred years old, depending on how quickly the ice is replenished. The ice caps atop mountains also flow, but because they are flatter, they flow much more slowly. And because mountain ice caps are found at the very highest elevations, they don't melt much; it's just too cold. This means that the ice within them is often much older than the ice within alpine glaciers. For example: A team of scientists drilled a record-breaking 1073-foot-long ice core atop Mount Logan in Canada's Yukon territory in 2022. The core is estimated to contain thirty-thousand-year-old ice. The France-based Ice Memory Foundation is trying to core and save some of this precious ice before it disappears.

The Ice Memory Foundation was established in 2021 to collect and store ice cores from glaciers most at risk of disappearing in the near term. With partners in Italy, France, and Switzerland, scientists have collected ice cores from Svalbard, the Alps, Russia's Carpathian Mountains, and Bolivia. For the time being, these ice cores are being stored in an ice cave in Antarctica. By the time you read this, ice cores will likely have been collected from several other glaciers, including those in the Pamir Mountains and Himalayas.

By preserving cores from mountain glaciers around the world, the Ice Memory Foundation is preserving more than just ice; they are preserving slices of environmental and human history. Our understanding of past cultures is incomplete without an understanding of past climate.

While climate doesn't determine our future, it has shaped and will continue to shape our lives in meaningful ways, even as we continue to shape the climate.

If we reach a point when all the ice left in the world is stored in dark warehouses, all we will have are our memories of wild ice. Future generations will only have photos and documentaries to remind them of what once was; they will never get to experience wild ice for themselves. Not only will they never be awestruck by a glacier calving into the ocean or an iceberg towering above them, but they also won't get to enjoy the simple pleasures of winter that many of us have engaged in during our own lifetimes, such as ice skating, playing pond hockey, careening downhill on a sled, or building a snowman. These simple pleasures are, of course, dwarfed by the magnitude of loss that Arctic and mountain cultures are experiencing. Their very livelihoods and cultural traditions are imperiled. Yet, we must not dwell in loss.

National Geographic Explorer and author Dr. M Jackson writes that to gaze upon a glacier is to be in both the past and the future. We imagine a past when there was more ice and a future in which there is no glacier left. Yet if we expect total loss of ice as inevitable, we lose hope, and inaction follows loss of hope. It's true that we face many challenges when it comes to climate change, from extreme weather events, extended droughts, and heat waves to food shortages and displacement of communities—in some cases, displacement of entire nations. But global apocalyptic generalizations of our future overshadow the possibility of human adaptation and resilience. Adaptability is built into our very DNA. It's time we harness it once again to adapt to the changes that are inevitable while also adopting a new way of interacting with the planet that won't contribute to the loss of ice and a hotter planet that may soon be outside the physiological range of human adaptation.

When the last ice age ended 11,700 years ago and the vast continental ice sheets began to retreat, life followed close behind. As plants colonized the land, animals followed, and we followed the animals. Not only did humans go north, but our ancestors also went up into the mountains as glaciers retreated to the highest elevations. It seems from the extent of archaeological evidence in the alpine, extending to the upper paleolithic, that humans immediately recognized the economic importance of alpine environments as soon as the ice retreated. We turn next to an alpine pass in Switzerland where archaeologists have uncovered seven thousand years of human history buried in a single patch of ice.

FOUR

Archaeology with Altitude

In Switzerland's Bernese Alps, between the headwaters of the Rhône river and the valley of the rivers Simme and Kander, lies a narrow notch eroded out of hard alpine limestone. At 9042 feet, this pass—known as the Schnidejoch—is well above the timberline, a landscape of rock, scree, ice, and wind. Barely two hundred feet across, the pass connects the summits of Schnidehorn and Wildhorn, two in a chain of crenellated peaks that tower above the valleys like watchmen on never-ending patrol. It is in this notch that 6800 years of history melted out of a withering patch of ice.

The discovery of the Schnidejoch ice patch began, like so many ice-patch sites, with hikers. On September 17, 2003, Ursula and Ruedi Leuenberger had stopped at the pass to rest and to admire the views of the Bernese Alps. They had slept at the nearby Wildhorn Hut of the Swiss Alpine Club the night before and were headed to another hut on the other side of the pass when they found a strip of birch bark folded over like a sleeve. The birch bark had been lying in a puddle of melt-water at the edge of the ice. It was about a foot long and sopping wet. Ursula noticed a series of holes along one side of the birch bark, as though it had been sewn together to create some sort of container. For what, Ursula couldn't be sure.

Although Ursula couldn't identify the object, she suspected that the birch-bark sleeve was more than it seemed. Rather than risk damaging it by stuffing it into her small backpack, Ursula carried the mass of wet bark in her hands across the Alpine ridge to the Wildstrubel Hut about four hours away. The hut proprietor gave her a small box, in which she carefully placed the object, then strapped the box to her backpack for the hike out the following morning. When she arrived back in Bern, Ursula delivered the box to the Historical Museum. A month passed before the object made its way to Albert Hafner, an archaeologist who at the time worked for the Archaeological Service of Bern.

The summer of 2003 had been dangerously hot. According to Switzerland's Federal Office of Meteorology and Climatology, June through August were the warmest months observed in the previous 250 years. Switzerland's hottest day occurred on August 11. Thermometers read 107°F (42°C). Weeks of unbearable heat caused exceptional melt. Across the Alps, glaciers lost almost ten feet of snow-water equivalent (the amount of water contained in snowpack if it were to all melt instantaneously), which was near twice the loss logged during the previous record set in 1998. That's a loss of nearly 4 percent of all the ice in Switzerland, which sounds like a small number, but it isn't. Imagine the amount of water it would take to fill 1,160,000 Olympic-sized swimming pools. It was a staggering amount of melting.

Extreme heat combined with drought led to widespread crop failures, many large wildfires, and record low flows in rivers. Swiss farmers were forced to import livestock fodder from other countries that were less affected by the severe climate. Water bans became commonplace. France had trouble cooling their nuclear reactors without river water. Around 80 percent of France's electricity needs are met by the country's nineteen nuclear power stations. At least four reactors were forced to shut down to avoid catastrophe, yet the demand for energy soared as people struggled to cool their homes. All told an estimated thirty thousand Europeans died of heat-related causes that summer. Half of them lived in France. According to the Intergovernmental Panel on Climate Change (IPCC)—the world's foremost authority on climate change—heat waves like this are expected to occur every other year by 2080. Perhaps we have already reached that dreadful mark. As it stands now, extreme events like this have occurred in six of the last twenty years, including 2017, 2018, 2022, and 2023.

Those who could escape the cities fled to the Alps in hopes of finding relief in the mountains. But the mountains were not without their own dangers. Ninety people were evacuated off the Matterhorn when melting permafrost released a mass of rock and debris, nearly killing everyone in its wake. Similar dangers on Mont Blanc forced climbers off the mountain. Hut proprietors warned hikers of dangerous avalanches, falling rocks, and flash floods, not to mention the potentially deadly consequences of plain old heat exhaustion.

By the time the Leuenbergers visited the Alps in September, the blistering heat wave had passed. Just two days after Ursula picked up the birch-bark sleeve, another group of hikers, among them Germans Bernhard Wolters and Hartmut Korthals, were crossing the same pass when they found a wooden stick a little more than five feet long. The carved stick tapered at both ends like a crescent moon—a longbow like the one Ötzi carried. Several shorter, wooden arrow shafts, also carved but straight rather than curved, lay about. It was obvious to the hikers that these were artifacts, but whether they knew that they had found an ancient bow with arrows is uncertain. They took these objects home to Wiesbaden, Germany.

The Mysterious Birch-Bark Sleeve

Meanwhile, Hafner puzzled over the birch-bark sleeve. "When we first saw the object, it was absolutely clear that it must be human-made," Hafner told me. "I thought it might be a quiver to hold arrows," he said. But he only had one piece of it. The remaining pieces were still on the mountain, hidden beneath the ice and awaiting their turn to melt out. Since it would be a year before anyone could get back up the mountain and look for the missing pieces, Hafner did what any archaeologist would do with an organic artifact—he submitted a sample for radiocarbon dating. But the date couldn't tell him what the birch-bark sleeve was or how it had been used, so he set it aside for the time being.

In plain old dig-in-the-dirt archaeology, artifacts are buried in more or less chronologic order in the ground. The deeper the layer, the older it is, although sometimes the layers get mixed up due to erosion, the digging and tunneling of rodents, and our own activities like mining and farming. But one of the characteristics that separates ice-patch archaeology from more traditionally understood, soil-based archaeology is that there is no excavation. By the time ice-patch artifacts are discovered, they've already been removed from their original depositional positions. Ice-patch artifacts are simply found on the surface of the ice or along its melting edge—kind of like they magically appeared, gifts from the ice.

Of course, Hafner knew about Ötzi the ice mummy, but that was, as far as he knew, a one-off discovery. American archaeologist E. James Dixon told me that the magical-seeming appearances of ice-patch artifacts has been a point of criticism among traditionalists in the field, a critique he says is moot, because so many of the objects are organic and therefore can be radiocarbon dated, a process that is far more precise than relative dating methods used at dig sites without organic remains. Ice is the world's best natural preservative. Freezing prevents the growth of microorganisms and fungi that feed on organic material. In soils, microorganisms and fungi thrive.

Besides, there is no getting around the fact that ice-patch artifacts are not found in tidy layers—for example, several thousand years of ice can melt all at once, leaving artifacts from very different time periods sitting next to each other on the surface, as if they existed in the past at the same time. The remarkable preservation of ice patches and the organic nature of the artifacts the ice preserves is part of what is so fascinating about ice-patch discoveries. Furthermore, what these artifacts lack in chronological organization, they more than make up for in individuality. Many ice-patch artifacts are unique, never-before-seen objects that require a bit of sleuthing to figure out what they were and how they were used. The birch-bark sleeve would turn out to be one such artifact.

A few weeks later, Hafner received the results of the radiocarbon analysis. The birch-bark sleeve was 4700 years old, placing it within the late Neolithic, about six hundred years after Ötzi's time. This major discovery had been just sitting there, in a puddle, like it fell off someone's backpack just days earlier, not thousands of years earlier.

Solving the Mystery

On a late summer day the following year, Hafner's colleague Kathrin Glauser, a scientific illustrator and archaeologist in Bern, climbed to Schnidejoch with the Leuenbergers. To get to the pass, they hiked seven miles, ascending more than 5500 feet in altitude. The hike took the better part of a day, so they stayed the night at the Wildhorn Hut. The following morning, they hiked along a knife ridge of scree at the northern flank of the Tungel Glacier. After 3500 feet of additional altitude gain, they reached the Schnidejoch.

By late summer, most of the seasonal snow in the mountains has melted, and glaciers and ice patches are at their smallest. This makes late summer and early autumn particularly good times of year for finding ice-patch artifacts. As Glauser carefully documented the site, noting the size of the ice patch and the surrounding landscape, Ursula wandered off to have a look around. She turned over a few rocks with the toe of her boot. Pools of meltwater gathered in the depressions left behind. She stayed vigilant despite finding nothing.

Searching for artifacts can be addictive. The possibility of finding something is almost (but not quite) as thrilling as actually finding something. Ursula hoped that the birch-bark sleeve wouldn't be her only discovery, but she tried to be realistic. Two discoveries would be unusual for someone not trained in archaeology; still, she kept search-ing. She took a few more steps and froze. Something that clearly was not a rock lay in the dirt. She picked it up and brushed away the soil, revealing a pin about nine inches long.

The pin had a large, flat disc at one end. A thin patina of oxidized bronze gave it an antique-emerald finish. Two concentric circles enclos-ing a central cross made up of eight chevrons decorated the disc head. She would later discover that pins of this style date to the Early Bronze Age (3700 years ago) and were worn by both men and women to fas-ten a cloak or robe. Most often they are found in burial sites, but this pin must have fallen off while whoever wore it was crossing the pass. These two objects—the Neolithic birch-bark sleeve and the Bronze Age pin—were separated in time by a thousand years but found in the exact same spot. With these two discoveries, Hafner knew he was onto some-thing, and the first ice-patch archaeology project in the Alps was born.

That autumn Hafner put out a news release about the Schnidejoch pass. Although he winced at the thought of tipping off the public that this was a site where artifacts could be found melting out of the ice, dozens (maybe hundreds) of people were already hiking across the site every summer. He figured that people might have already taken objects from the pass. While a news release would alert the public to the location of a previously unknown archaeological site, Hafner also hoped to educate hikers on the importance of leaving artifacts in their original position and alerting local archaeologists. It was a risky move but one that paid off.

On the evening of November 11, 2005, the hiker Bernhard Wolters was sitting in his living room watching the German news channel ARD when a reporter began talking about the Schnidejoch and artifacts that had been found there. The news release urged anyone who had taken home objects collected at the pass to turn them in. A phone number for the Archaeological Service in Bern flashed across the screen. Remembering the wooden sticks he and Hartmut had taken from the site, Bernhard immediately called the number. A month later, Glauser traveled to Germany to collect the bow and arrows. "I suppose we could have fined them," Hafner told me, "but we were just glad to have them back." These were the only artifacts turned in after the news release, but they were key clues in helping Hafner identify the birch-bark sleeve.

Over the next couple of years, Hafner's team found the two missing pieces of the birch-bark sleeve—a two-and-a-half-foot-long straight sleeve open at both ends and a smaller sleeve that curved toward one end that had been sewn shut. All three parts were made of the same materials and constructed in the same way, so Hafner guessed they must be parts of the same object. Hafner easily fit them together. When fully assembled, the birch-bark sleeve was just long enough to fit the unstrung bow. Not only did the sleeve fit the bow, but it was tapered along one side of the case to accommodate the bow's convex shape. The birch-bark sleeve was a Neolithic bow case. The case is the first and only one of its kind ever found.

The sleeve was made of around forty strips of birch bark that had been placed in an overlapping pattern like shingles on a roof, allowing water to easily slide off without soaking the contents. The shingles were sewn together with plant fibers. A second layer of bark ran lengthwise down the inside of the bow case. The whole thing was reinforced with two long wooden rods that ran the length of the sleeve. The bow would have slid into the sleeve from the top, which was then capped with the section of birch bark Ursula had found two years prior. A similarly sized rawhide piece fit on the bottom part of the sleeve, which remarkably still held two small stone points. A leather strap attached to the rods allowed it to be slung over the owner's shoulder.

Although the find is a rare one in the archaeological record, it's hard not to see the utility of a bow case. It would have freed the Neolithic hunter's hands when they were not actively hunting while also protecting the bow, especially from moisture. A saturated wooden bow changes shape, and firing a soft bow would ruin it forever.

Since the bow, arrow shafts, and birch-bark case all dated to the same age, Hafner assumed they had all belonged to the same person. In just a few years, Hafner had collected a complete Neolithic bow kit similar to the one Ötzi had carried, and he speculated that the bow kit had not been forgotten or misplaced by its owner. Hunting tools like these would have been highly prized, especially the bow. A hunter expected to frequently replace arrows, but he or she had only one bow, which took at least a full day to carve and at least three weeks to dry once cut and shaped. When Hafner found a long piece of leather that may have been part of a legging, a pair of leather shoes, and a grass cape that all dated to the same age as the bow kit, Hafner suspected or perhaps even hoped that another ice mummy like Ötzi lay buried in the ice somewhere nearby.

In Search of Another Ice Mummy

To find out if there might be a body near the pass, Hafner hired a team of cadaver dogs in early autumn 2006. The dogs were trained on bones from the Roman Age to help them identify the scent of old remains, but the case was probably a bit too cold. The only bones the dogs scented were those of animals. Glauser writes in 2015 that the "dogs played more than they worked." As it turns out, to a dog, a bone is still just a bone. Still, Hafner is convinced that the owner of the bow kit and clothing must have perished nearby. The Neolithic bow kit and clothing were such an unusual assemblage that it's unlikely these items were misplaced.

Despite its failure to produce a body, the Schnidejoch pass turned out to be extraordinarily rich in cultural material. During the eight years in which Hafner surveyed the ice patch, he and his team found nearly nine hundred artifacts spanning the last 6800 years. Most of these artifacts were organic—leather, wood, plant fibers, and bark. The oldest object from the Schnidejoch is a bowl made of elm wood dated to about a thousand years before Ötzi's time. You can just imagine the bowl's owner,

THE AGE OF MELT

having climbed from Valais in the valley of the Rhône river, stopping for a rest just after cresting the pass. Perhaps the bowl contained a meal like the one Ötzi would have enjoyed in his day: einkorn and dried ibex or salted fish, maybe. Resting at the top of the pass, she might have gazed back toward her village in the valley below, barely visible as a collection of small farms, with gray clouds rising from thatch-roofed huts. Perhaps the bowl broke and she tossed it onto the snow, or maybe she lost it in just the right place for fragments of it to survive millennia. Whatever the story is behind the bowl, we now only have its pieces as evidence of the life of its owner.

Despite the absence of human remains at Schnidejoch, the objects that had melted out of the ice were made by people just like us, and these belongings were important enough for their owners to carry them up and over the Alpine ridge. But why? Who were these people and where were they going?

A Walk Between Life and Death

The Alps divide Northern Europe from Southern Europe in a discontinuous 750-mile-long chain spanning eight countries. At their broadest, the Alps are 125 miles wide between Garmisch–Partenkirchen, Germany, and Verona, Italy. It could take weeks, maybe months, to walk around these portions of the range. So rather than embarking on weeks-long journeys to go around the mountains, people traveled through them, choosing relatively low-elevation passes that led them from one valley to another. Though far faster than circumnavigating them, traveling through the mountains was not easy by any stretch of the imagination and presented travelers with many dangers, both real and perceived.

A dozen or so mountain passes in the Alps are known to have been used as far back as the Roman Era. By the first half of the nineteenth century, roads had been constructed across many of the most well-traveled passes. One of the most famous of these Alpine crossings is Switzerland's Saint Gotthard Pass. The pass became popular after about 1220 CE, when a wooden bridge was constructed over the narrow, steep-sided Schöllenen Gorge above the Reuss river. As the Reuss was turbulent and swollen with snowmelt during the early summer, the bridge saved people from a treacherous ford. But crossing the pass was still difficult, even centuries later. Of his own crossing in the late eighteenth century, German poet and playwright Friedrich Schiller wrote:

> *You walk between life and death. Two threatening peaks shut*
> *in the solitary road. Traverse noiselessly this place of terror;*
> *fear to awaken the sleeping avalanche. The bridge which*
> *crosses the frightful abyss, no man would have dared to build.*
> *Below, without power to shake it, growls and foams the torrent.*
> *A sombre arch seems to conduct to the empire of the dead.*

The most renowned Alpine crossing was made by the Carthaginian general Hannibal in 218 BCE. While the exact pass Hannibal crossed eludes historians, ancient texts reveal that he led tens of thousands of foot soldiers, thousands of horses and mules, and thirty-seven African elephants on a fifteen-day journey through the mountains. His goal was to destroy the Roman Republic, making the bold move to cross the Alps in mid-autumn when snow could fall at any moment. Although Hannibal lost roughly half his troops (and an unknown number of elephants) during the passage, the Romans mistakenly believed the Alps to be a barrier. Hannibal's army raged against the Romans for the next fifteen years. Although Hannibal enjoyed many victories, the Romans eventually won the war. Nevertheless, Hannibal's crossing is considered one of history's greatest military achievements.

Although Hannibal proved the viability of crossing the Alps, those who lived among the mountains were not under Roman control, which was a problem because the mountains separated the various parts of the Roman Empire from one another. Romans who traveled through

87

the mountains were often attacked and robbed by one of the forty-six Indigenous Alpine tribes who already lived there. To remedy their lack of control of Alpine tribes, armies marched across the mountains in a centuries-long battle to conquer them. After the Romans assumed control of the Alps by the sixth century BCE, they built roads, monasteries, hostels, and hospices at key passes to provide safe passage and medical care to travelers.

The Great Saint Bernard Pass (of dog-breed fame) in Switzerland is one such pass. It's the same pass that French military commander Napoléon Bonaparte victoriously crossed with an army of forty thousand soldiers in late spring 1800 in his bid against the Austrian army in Italy. Great Saint Bernard Pass lies between the two highest mountains in the Alps—Mont Blanc and Monte Rosa. The pass itself lies at about 8100 feet. In 1857, Danish journeyman Theodore Nielsen wrote of the difficulties in crossing the Great Saint Bernard Pass. Nielsen spent the night in the Saint Bernard monastery and hospice after a long climb to the saddle.

> *We arose at five the next morning and were given a very good and generous breakfast before we started our descent of the mountain. The clouds were threatening, black and so heavy that we could see nothing and wished earnestly that we were back onto the green earth once more. It was hard work going up the mountain but worse going down. We sank into the snow, several times so deep that we had all we could do to get up again.*

Traditional Land Use in the Alps

While passes like the Great Saint Bernard are well recorded in history books, the Schnidejoch was unknown until artifacts began melting out of the ice. For thousands of years before the Schnidejoch became a recreational hiking trail, the pass served as a travel corridor connecting the villages of Lenk in the Bernese Oberland to Ayent in the Valais. As if to underscore its importance to travelers, one of the most common objects Hafner's team found were Roman-age hobnails used to secure soles to the leather uppers of shoes. Not only did the hobnails hint at the number of travelers—at least during the Roman Age—their number also showed that shoes often fell apart (one of the other items Hafner's team found was a shoe-repair kit).

With nine hundred artifacts spanning 6800 years, Hafner began looking for patterns in the chronology of the objects. He found that the radiocarbon dates clustered around three periods that were separated by hundreds or thousands of years for which there were no finds. The first cluster of artifacts dated to the late Neolithic, between 6800 to 4400 years ago. The second cluster dated to the Early Bronze Age, from about 4100 to 3500 years ago, followed by a dearth of artifacts for the next 1600 years. A third cluster includes artifacts from the Iron Age, Roman Age, and Middle Ages, dating to about 2200 to 1000 years ago.

Hafner wondered whether the clustered dates of the artifacts could have something to do with climate. Although modern travelers can easily cross the pass today, the ice patch on the north side had once been part of the Tungel Glacier. The Tungel Glacier lies to the northwest of the saddle. The current trail leads along a gentle slope north of the ice patch down to the front of the glacier. When the Tungel Glacier advanced, it would have made crossing the pass impossible. To find out if the ages of the artifact clusters had something to do with the advance and retreat of the glacier, Hafner teamed up with Martin Grosjean, a paleoclimatologist with the Oeschger Centre for Climate Change Research in Bern. Although Grosjean is now the Director of the Oeschger Centre, he still makes time for fieldwork, especially when it involves getting into the mountains.

While scientists have been monitoring European glaciers since the Little Ice Age, no one was recording glacier length during the Neolithic. To map out periods of advance and retreat, Grosjean reviewed published studies detailing glacier change at nearby sites. These studies relied on rooted tree stumps that had been killed by advancing ice. By counting tree rings, scientists are able to determine the minimum length of time during which glaciers were in retreat and the tree was growing. The radiocarbon dates of rooted trees indicate when the tree died and when the glaciers expanded.

Grosjean found that the breaks in dated artifacts from the Schnidejoch corresponded to periods of glacial advance. In colder periods, the Tungel Glacier flowed down the Wildhorn, preventing travel across the pass. The very narrowness of the Schnidejoch pass means that it would have been very sensitive to glacial changes. Up until the 1990s, the Tungel Glacier was impassable without specialized mountaineering equipment. Photos from the 1920s show deep crevasses as the glacier slid over a steep drop. As Grosjean writes, the "Schnidejoch is a binary and non-continuous archive." It is either open or closed to travelers.

The enormous volume of artifacts coupled with the long span (about seven thousand years) prompted paleoecologist Christoph Schwörer at the University of Bern to link the archaeological finds from the Schnidejoch pass with paleoecological data. (Paleoecology is the study of ancient environments and how they have changed through time.) Schwörer wanted to find out not only how plant life shifted through time, but what role humans had in that shift.

In 2010, Schwörer pulled a twenty-five-foot-long sediment core from the center of Lake Iffigsee, which lies on the south side of Schnidejoch pass some two miles northeast and 2300 feet lower. Today, the turquoise lake is surrounded by green alpine meadows with an occasional larch tree providing a bit of shade to weary hikers. It's an idyllic spot, but it hasn't always looked this way.

When the Pleistocene ended 11,700 years ago, Lake Iffigsee, which was no more than a deep hole carved by glaciers, filled with meltwater. Over time, soils developed around the lake and plants took root. As these plants shed pollen, pine needles, and leaves, some of this organic matter blew onto the lake and sank below the surface, collecting in dateable chronological layers on the lake bottom, much like glaciers accumulate annual layers of snow. The low-oxygen environment of the lake bottom prevented the growth of decomposers, and the plants were preserved so well that many of them can be identified to species.

While lightweight pollen travels long distances, recording plant species on a regional scale, heavier plant parts like leaves and needles can be used to reconstruct the plant community on a hyperlocal scale. The types of plants (such as trees, grasses, and shrubs) and species (such as pine versus birch) not only reveal what the plant community was like in the past, but they also correlate to past climates, because of course, different plant species are adapted to different climates.

Schwörer found that for most of the Holocene, Lake Iffigsee was surrounded by forest. As the Pleistocene came to a close, temperatures climbed. Within the first half century to a century, the temperature rose by about 3.6°F (2.0°C). Warmer temperatures allowed trees to colonize the area. When temperatures peaked around 9800 years ago, the climate had warmed to about 2.7°F (1.5°C) above today's averages. It's a period called the Mid-Holocene Warm Period, and it lasted about four thousand years. During this time, people settled the Rhône valley. While they weren't yet farming in the Alpine, people still traveled into the mountains to hunt ibex, chamois, and other animals for their furs and meat.

Beginning about seven thousand years ago, the amount of charcoal in the lake core spiked. The regularity of the signal suggested that people were using fire to clear land for pastures. Fire has the added benefit of stimulating the growth of nutrient-rich grasses for grazing livestock. Schwörer also found an increase in the spores of a fungus that only grows on animal dung. While *Sporormiella* is found naturally in the environment, large concentrations bring to mind livestock corralled in some kind of enclosure or pen. The added nutrients supplied by cattle dung fertilized the soils, and grazing removed existing plants, making way for disturbance-tolerant and nutrient-loving plants like stinging nettle and sorrel. These Neolithic pastoralists probably came from villages in the

Rhône valley on the south side—a twelve-mile walk from the river to the Schnidejoch. With goats or sheep, the trek probably took a day or two.

Transhumance is a type of pastoralism in which herders drive livestock in seasonal rounds between winter areas and summer areas. In the Alps, these routes range from low-elevation wintering grounds to high-elevation summer pastures. The widespread and historical practice of transhumance in the Alps is why Spindler's Ötzi-as-shepherd story made so much sense.

In early summer, the Ötztal Alps ring with sheep bells and the "höörla leck leck leck" calls of shepherds as they drive their flocks up and over the main Alpine ridge to graze in the lush green valleys of Vent in Austria. The twenty-seven-mile trip takes two days, but herders constantly move sheep throughout the summer to find the best grazing and lambing pastures. In early autumn, the route is reversed. But a detailed analysis of plants that thrive alongside livestock showed that transhumance in the Ötztal Alps didn't begin until a full 2200 years after Ötzi's time. This debunked the Ötzi-as-shepherd theory. At Lake Iffigsee, however, Schwörer found that people were pasturing livestock three thousand years earlier than where Ötzi lived. It's hard to say why there is this difference in timing. Perhaps it had to do with differences in local climate or population size. One of the reasons people may have begun pasturing at higher elevations is that as populations rose, there just wasn't much room in narrow valleys for both people and animals.

Hafner and his team found particularly intriguing evidence of pasturing in the mountains around Schnidejoch: a collection of five separate plaited rings made of birch and spruce twigs. Four rings, each six inches in diameter, all date to the early Bronze Age. The fifth dates to the late Iron Age. At first, Hafner thought that the rings were used to secure items to pack animals, but he soon discovered that similar wooden rings are still being used in mobile fencing systems in the Alps, including in the valleys below Schnidejoch.

Hafner told me that once when he gave a talk to local residents of the Rhône valley—before he knew how the rings were used—an older woman greeted him afterward. She told him how the rings had been used to secure fences. Although she hadn't used the same type of rings herself, they were heirlooms that decorated the walls of her cabin. Fences would have been important to protect livestock from nocturnal predators like bears and wolves and to keep livestock from wandering. Although Hafner doesn't know quite what the fences looked like thousands of years ago, today's fences, known as ringzaun, or "ring fence," are made of canted wood planks tied to vertical sticks with the same style of plaited rings found at Schnidejoch.

From the Iron Age onward, Schwörer found a significant increase in cereal pollen, such as from barley. Bringing livestock to high alpine pastures might have been what encouraged people to start summer farms in the Alpine beginning in the Iron Age, although they probably weren't farming right around the lake but at lower elevations. During periods of low-intensity land use, forests recovered up until the Middle Ages, when timberline was pushed below the lake, leaving behind the alpine meadows we are familiar with today.

Because these types of traditional land uses are still practiced in many parts of the Alps, they might mitigate some of the effects of climate change on sensitive Alpine vegetation by preventing the establishment of trees higher up the mountain as the climate continues to heat up. In other places, where traditional grazing practices have been abandoned, especially on particularly steep slopes, trees are beginning to grow. This natural regrowth of forest has encouraged wolves, bears, and eagles to return to some areas.

The Forgotten Pass

The Schnidejoch pass fell out of favor sometime during the early Middle Ages, just as the Little Ice Age began. Throughout the Little Ice Age, the Tungel Glacier extended about four thousand feet farther down valley than it does today. The glacier blocked the pass until sometime after 1850. But not all passes were blocked during the Little Ice Age, and some were counterintuitively easier to cross when filled with ice, especially during winter. According to a 1652 report, the Lötschen was one such pass. In the report, the author writes:

> From Kandersteg, two passes allow you to pass through the mountains into Valais. The first is the Gemmi path and the other the Gasterntal and Lötschenberg path. We know of this pass on the Lötschenberg that it is impassable in summer because of the numerous and deep crevasses in the mountain; but it can be used in winter, because the crevasses in question are filled with snow, and this snow freezes enough for the Italians to be able to cross it with cattle.

Glaciers straddle both sides of the pass—the Langgletscher to the west and the Grosser Aletschfirn to the east. The latter is one of three icy tributaries that converge to form the largest glacier in the Alps—the Aletsch. It's hard to overstate the size of this glacier. From above, its three tongues converge to form a twenty-four-mile-long slithering mass of ice that curves to the west, growing skinny as it reaches its terminus 8200 feet lower than its point of origin. From above, one gets a sense of the glacier's movement. Two dark ribbons of rock (which are actually medial moraines that form when two glaciers come together) stream in the direction of glacier flow.

Aside from this brief report, the Lötschen Pass would have been forgotten had it not been for Swiss painter and photographer Albert Nyfeler, who often camped near the pass to paint the glaciers during the 1930s and 1940s. After completing his artistic training, he built a house in the Lötschental valley in a village called Kippel. Of the pass he wrote:

> The Lötschental offers me everything ... water, trees, and huts, animals, people, mountains with the glistening glaciers and

94

the blue sky, pure color and dark like nowhere else. I couldn't imagine that I would live and work in a city forever today. I would always be drawn to the mountains.

Over the years, Nyfeler collected ancient bow fragments, medieval crossbow bolts, arrow shafts, and pieces of leather. These items all dated to the end of the Neolithic and the Early Bronze Age, which is about 4400 to 3800 years before the present. Although archaeologists have visited the site since the artifacts were recovered from Nyfeler's workshop in 1989, no other finds have been recovered there since.

* * *

During the first few years of the Schnidejoch ice-patch project, Hafner estimated that the ice patch had lost roughly half its area. He expected the ice patch to disappear altogether within a year or two, but against all odds, the ice patch persisted. And then the summers of 2006 through 2008 were cool enough that the ice patch stabilized. But once again, a year later in 2009, the ice patch melted back considerably. While the ice patch had persisted beyond Hafner's expectations, he could see that it would eventually disappear. It was only a matter of time. And then at summer's end in 2018, the Schnidejoch ice patch finally melted completely, leaving behind a collection of loose boulders and a few remaining artifacts. "It was a long agony," Hafner told me.

Prior to the emergence of ice-patch archaeology, archaeologists assumed that most mountain ranges—especially those that support glaciers and perennial ice patches—were inhospitable, loathsome places full of topographic puzzles no human had attempted to solve until recently, and then only as a matter of conquest. In 1938, anthropologist Julian Steward wrote of the Great Basin Mountain ranges in North America as "unimportant to man, except as [they support] animal species." Alfred Kroeber echoed this sentiment in 1939 when he wrote, "like other elevated divisions, the Rocky Mountains constituted chiefly fringes, hinterlands, or barriers under native settlement. There was no ... pressure ... to draw the population into the mountains." And George Bird Grinnell, a conservationist who played a leading role in establishing Glacier National Park, which is the ancestral home of the Blackfeet, Salish, Pend d'Oreille, and Kootenai, described the area as "absolutely virgin ground ... with no sign of previous passage," even as he relied on the Indigenous Blackfeet to guide him over traditional pathways through those same mountains.

While people have been crossing snow and ice in the Alps for as long as it has been feasible, ice itself wasn't the draw. These bodies of ice were encountered by chance as people traveled across them, losing or tossing belongings that would one day be considered artifacts. No one crossing the Schnidejoch could have imagined that their loose change, old shoes, and broken arrows would one day be part of museum collections, the subject of documentaries, or written about in books.

Although mountain passes have provided people with avenues of travel for thousands of years, there were many other reasons people spent time in the mountains. Hunting, for instance. Ötzi and whoever lost their bow atop Schnidejoch hunted chamois, ibex, bear, and other animals for their food and furs. Crystal seekers also spent time in the mountains searching for materials to make hunting tools. Later, during the Copper and Bronze ages, miners headed into the Alps for metals. But ice wasn't the draw for those who ventured into the Alps; ice was simply encountered in passing. However, in Norway, archaeologists have uncovered a unique reindeer-hunting tradition that drew hunters specifically to the ice.

FIVE

Digging into Norway's Ice

Mimisbrunnr Klimapark in south-central Norway is unlike any other park in the world. Part museum, part art exhibit, Mimisbrunnr is a 230-foot-long tunnel carved into the Juvfonne ice patch at the edge of Jotunheimen National Park—the land of the frost giants (the *jötnar*) of Norse mythology. The land of giants is said to be filled with mountain peaks where winter never eases its grip. According to legend, the frost giants were the first creatures to inhabit the world, born of fire and ice in the great void at the beginning of time.

Mimisbrunnr is named for the Norse water spirit Mimir, who guards the well ("brunn") of knowledge that lies beneath the world tree Yggdrasil. Mimir's waters are so powerful that Odin traded an eye for a draft of its liquid wisdom. But Mimisbrunnr wasn't created to commemorate stories in Norse mythology—at least not entirely. The park was founded in 2010 after relics belonging to ancient reindeer hunters began melting out of Juvfonne and other nearby ice patches.

The ice tunnel was excavated in 2010 to showcase the enduring relationship between nature, climate, and humans through time, including our role in the current climate crisis. As if the tunnel itself understood its mission, it collapsed the following summer when June and July temperatures were 3.6°F (2.0°C) warmer than the average for the previous decade. Not to be deterred, Mimisbrunnr Klimapark's founders rebuilt the tunnel in 2012. Ever since, the ice has withstood progressively warmer summers. The summer I visited, however, was cold and snowy.

I visited Mimisbrunnr during a snowstorm. It was the first week in July. Temperatures hovered just above freezing, but with the wind chill it felt more like 20°F (-6.7°C). At the 62nd parallel north and 6040 feet in elevation, I should have expected colder weather, but this particular year had been unusually cold. "June was just shit," said Ingrid, the guide for our small group of visitors on the day I toured Mimisbrunnr. For a Norwegian to say the weather was shitty says a lot about the kind of spring they'd had. July wasn't shaping up to be any better.

I squinted against the sting of ice whipped up by gusts of wind blowing at more than thirty miles per hour. Icebergs floated along the margin of Juvvatnet, a small, glacier-fed lake at the edge of the Klimapark. Snow buntings atop the bergy bits foraged for insects immobilized by the cold. Luckily, I had read the weather report before taking the bus up the mountain from Lom, a small town situated at the confluence of the Bøvra and Otta rivers, in the foothills. The Bøvra river, barely contained by steep granite walls, charges through town in torrents of white foam and roiling turquoise water. Cold mist rising from the river beaded on my rain jacket, though it was hard to tell which soaked me more—the mist from the river or the days of rain. Although the mountains were invisible beneath a cloak of clouds, I knew that when it rains in Lom, it snows in the mountains, so I purchased a cheap down coat and wind pants at a local shop filled with outdoor gear for people like me who hadn't come prepared for a midsummer snowstorm.

"Ah, look. There's the mountain," said Ingrid between gusts of wind. I turned to face Galdhøpiggen—the tallest mountain in all of Scandinavia, although it's only about 8100 feet above sea level. But what Scandinavia's mountains lack in height compared to other mountain ranges like the Alps, they make up for in ruggedness. These mountains are deceptively steep, rocky, glacier-covered, and wildly changeable.

Not long ago another peak called Glittertind, also in the Jotunheimen Mountains, held the coveted title of tallest peak. In 1917, the elevation of its glacier-capped summit rose to a height of 8137 feet. But a second survey in 1965 found that the ice atop its summit was melting, sparking debate about whether a glacier should be included in the height of a mountain. The point became moot in 2017 when Glittertind's glacier disappeared, revealing its bald granite summit, which stands just fifty-five feet shy of Galdhøpiggen's highest point. Although Galdhøpiggen's slopes are draped in glaciers, its peak is free of ice, forever securing its title as the tallest mountain in Scandinavia.

Ingrid led us from the parking lot of the Juvasshytta mountain lodge near the Klimapark to a boardwalk that floated above a carpet of loose boulders, which in turn lay atop a vast expanse of permafrost. The boardwalk ended at a set of white double doors that led us into the belly of the ice patch. As the frosted doors fell shut behind us, a powerful boom echoed through the tunnel. It was a sound that could easily be mistaken for the footsteps of giants if people still believed in such things. The doors shut out all light from the outside. I paused a moment to let my eyes adjust to the dimly lit hall. Although the temperature was as cold inside as it was outside, the absence of wind was a relief. I pulled back the hood of my down jacket and breathed in the deep, earthy aroma of millennia-old reindeer dung.

The Original Ice-Patch Archaeologists

In the exceptionally warm summers of 2002 and 2003, a few curious objects had melted out of one of Jotunheimen's ice patches. They were sticks about three feet long, each with a piece of birch bark attached to one end—different designs and seemingly different functions from the arrow shafts found in the Alps. They had been found lying atop a scree field in the forefield of an ice patch. These first finds foreshadowed an extreme melt in 2006 that would expose hundreds of ice-patch artifacts in a single year, including a shoe made of animal hide that was last worn more than three thousand years ago. The shoe would have fit a woman's size six or seven foot, or perhaps a teen boy. But these weren't the first ice-patch finds in Norway.

Long before the Leuenbergers picked up the bow case and the Simons stumbled upon Ötzi, a few dedicated amateur archaeologists and reindeer hunters in Norway had collected artifacts found along the edge of ice patches in the Oppdal mountains north of Jotunheimen. In 1914, a collector found an arrow—complete with fletching, sinew lashing, and projectile points—at the edge of a receding ice patch. This was the first published record of such a find and the first inkling of a deep-time relationship between humans, ice, and reindeer.

The collector offered the arrow and points to the Norwegian University of Science and Technology (NTNU) archaeology and cultural museum in Trondheim. In the museum's annual catalogue of newly acquired objects, the author of the entry makes a point of mentioning the size of the ice patch, of which he wrote that "it has not been this small for a long time." And then, beginning in 1936, a series of mild winters and extremely warm summers led to additional melt and more ice-patch finds.

In 1937, archaeologist Bjørn Hougen recounted in the Norwegian archaeological journal *Viking* the story of a hotel owner and his twelve-year-old son who found two arrows—one with an iron arrowhead and the other made of bone—while hiking up the steep north face of Storhø in the Dovrefjell mountains. Hougen writes: "[that] the arrows have been encased in ice or snow is certain enough, otherwise the shafts would have never held, but perhaps the strangest thing was that the arrow was stuck in the ground at an angle." Hougen was the first to recognize that the ice itself was the source of these arrows and that the ice had acted as a preservative.

Archaeologist Theodor Petersen wrote of these finds in 1951 that "this was something new, because I did not know of any other finds of this kind in this country [other] than the arrow that was discovered in 1914." Petersen went on to write that he "was immediately aware of the scientific importance of these finds, both for the history of hunting and for elucidating the climatic conditions in ancient times."

Aside from these brief museum catalogue entries and short notes in niche journals, few archaeologists in Norway had published much about these initial discoveries. Very little was known about what these artifacts meant and how they should be dated. The typology of these arrows wasn't well established at the time, and radiocarbon dating wouldn't become widely available until several decades after its 1946 discovery. As a result, these objects remained mere curiosities until the late 1960s, when Oddmunn Farbregd, an archaeologist with the NTNU museum in Trondheim, began to take an interest in this small collection.

Farbregd, also a beloved teacher of archaeology at NTNU, combed through the written materials on each object and began the task of classifying the arrowheads based on their shapes and the materials from which they were fashioned. His work was instrumental in developing the first typologies of arrows in Norway. You might say Farbregd was the original ice-patch archaeologist, quietly toiling away decades before Ötzi and the elm-wood bowl at Schnidejoch were discovered. He even coined the term "glacial archaeology" (ice-patch archaeology) in 1968 (although that term would be independently reintroduced many decades later by American archaeologist E. James Dixon).

Based on his review of the ice-patch finds, Farbregd concluded that nothing older than about 1900 years would be discovered in these ice patches because nothing older than that had melted out of them in all these years. It was a reasonable assumption. No one was talking about climate change at the time, and when Farbregd began looking at the museum collection in 1968, almost no new ice-patch artifacts had been found since 1943, despite regular surveys by a cadre of devoted collectors who volunteered for Farbregd. Nor would they find much for the next sixty years. Between 1944 and 2000, Farbregd's volunteers found just a dozen ice-patch artifacts. But a new era of ice-patch archaeology was about to begin.

A New Era of Ice-Patch Archaeology

Even though the Juvfonne ice patch is only a ten-minute walk from the Juvasshytta mountain lodge, no one knew that it held ice-patch artifacts until 2007, when amateur archaeologist and reindeer shepherd Jan Stokstad found several partially collapsed stone hunting blinds below its melting edge. As Stokstad continued searching, he noticed many wooden sticks caught in the mud and rubble left behind by the retreating ice. Recognizing the significance of these objects, Stokstad called local archaeologists with the Department of Cultural Heritage in Innlandet County.

Archaeologist Lars Pilø visited the site soon after. He and a few colleagues and volunteers found dozens of artifacts that year. By the end of the first three summers, they had found four hundred artifacts, almost all of which were the same type of curious wooden sticks found in 2002. But what got Pilø hooked on ice-patch archaeology was an arrow he found in 2006. The artifact was every archaeologist's dream—a perfectly preserved wooden arrow shaft with bird-feather fletching held to the shaft with sinew. "I found it so confusing because it didn't look old, but the arrow was clearly from the Viking Age," Pilø told me. The Viking Age lasted only a few hundred years, from about 800 CE to 1050 CE. It was an era of Scandinavian expansion owing to the development of the longship. Longships were designed to be light and fast, but most important, they did not require a harbor; longships could be pulled ashore, which gave the Vikings access to previously inaccessible shorelines from which to embark on their raids.

When Pilø encountered the complete arrow shaft, he wasn't an ice-patch archaeologist. His work had mostly been focused in the lowlands at the Viking Age town of Kaupang just south of Oslo, but from that moment forward, Pilø focused all his attention on surveying ice patches, eventually co-managing a research program and blog called *Secrets of the Ice*.

Almost all ice-patch artifacts are found in what Pilø calls the lichen-free zone. "The lichen-free zone is a halo of light-colored bare rock in the forefield of an ice patch. It shows where the ice had been in the recent past," Pilø told me over video chat. As an ice patch expands, it kills plants living in its wake just like advancing ice killed trees in the forefield of glaciers around Schnidejoch pass. Most artifacts are found in the lichen-free zone because any objects held within the ice as it melted back wound up on bare, lichen-free rock. And because organic artifacts begin to decay once exposed to sunlight and wind, they are usually only found in the forefield of recently exposed ice, where lichen hasn't had enough time to recolonize the rock. Any organic objects outside of this zone have long since decayed.

Lichen are slow-growing lifeforms composed of a moss and a fungus living in symbiosis. The fungus benefits from carbohydrates (sugars) produced by the moss, and in return the fungus provides nutrients, water, and protection to the moss. The species growing around Juvfonne is called map lichen (*Rhizocarpon geographicum*), which grows on rocks in cold mountain environments around the world. Its bright yellow-green center rimmed in black fungal hyphae requires more than twenty years to reach a diameter of about half an inch. By measuring the diameter of lichen on the rocks at the edge of the lichen-free zone, Pilø could not only map the size of the ice patch in the past, but he could also estimate when the ice patch had last been that large. It's a method of dating called lichenometry, which isn't anywhere near as precise as radiocarbon dating but would give Pilø an estimate of when Juvfonne had begun melting.

Based on the growth of the lichen, the Juvfonne ice patch had retreated more than 820 feet since its maximum extent during the Little Ice Age. Pilø also found that moss patches (without their fungal counterparts) were still rooted in the ground as the ice patch receded. By radiocarbon dating the dead moss, Pilø was able to figure out when the moss stopped growing, and hence when the ice patch had begun expanding. The moss was killed by an expanding ice field about two thousand years ago. In conjunction with aerial photos, this information led to Pilø establishing that most of the melt occurred during the last sixty years.

Out of the Ice

Being inside of the Juvfonne ice patch was nothing like being inside the Mer de Glace, which glistened with a thin veneer of meltwater, its cerulean walls seemingly radiating with light, and refreshing drops of glacial rain that occasionally splashed my neck. The inside of the Juvfonne ice patch was dark and shadowy. Its walls were rough to the touch, its ice cloudy, and there was the faint odor of decay.

One of the first things I noticed as my eyes adjusted to the light was a soft, orange glow midway down the tunnel. As I drew closer, I saw that the light was coming from a ring carved into the ice. Ingrid told us that the ring represented the Jörmungandr—the serpent of Midgard, the realm of men. In the Norse sagas, Odin—having heard rumors that Jörmungandr and its two siblings, a wolf named Fenrir and the goddess Hel, would one day rise against the gods—tossed Jörmungandr into the ocean in hopes that it would drown. But instead of drowning, Jörmungandr thrived in the sea, growing so large that it wrapped itself around Midgard, securing its hold by biting its own tail. It is said that when Jörmungandr releases its tail, the seas will flood the earth, setting in motion the final events of Ragnarok—a series of natural disasters that will lead to the end of the world.

The image of a serpent biting its own tail is one that appears in cultures around the world. It's known as an ouroboros. The ouroboros symbolizes rebirth, eternity, and the cycles of nature. But the ouroboros biting its own tail also suggests the dark side of self-destruction. The story of Ragnarok and the current climate crisis parallel one another. The world is burning. The seas are flooding the land. Natural disasters of our own making abound, from hurricanes in southern California to floods in Vermont and fires in Greece. In spring 2023, the skies around where I live in northern Wyoming dimmed with wildfire smoke from Canada. We're not used to seeing smoke-filled skies until late summer. The smoke reached the east coast, veiling the Manhattan skyline in a sickly orange fog. *This* is Ragnarok.

I passed under Jörmungandr and entered a large room filled with ice carved to look like the branches and twisted roots of Yggdrasil. Several tunnels split from the main room. I chose one and came face to face with an arrow and something called a "scaring stick." These replicas of the original artifacts were encased in a glassy window of ice lit with cool white light. The sign next to the exhibit told me that the original objects were about one thousand years old.

Unlike the diverse assemblage of artifacts found at Schnidejoch, almost all the artifacts found at Juvfonne were the same unadorned, three-foot-long sticks that had been found in 2002 at a nearby ice patch. Pilø figured that the sticks, the hunting blinds, and the ice patch must be linked, but how? Part of the answer appears as an illustration in an old book written in the eighteenth century.

In 1721, a Lutheran missionary of Dutch and Norwegian descent, Hans Egede, traveled to Greenland with his wife, four children, and forty other colonists. Egede lived in Greenland for the next fifteen years before returning to Denmark in 1736 with the body of his wife Gertrud, who died of smallpox the year before. Following Egede's return to Europe, he published an expanded version of an earlier book he wrote about Greenland, simply titled *A Description of Greenland*. The book included an illustration of how the Inuit hunted reindeer. It shows a line of people clapping their hands behind a small group of animals. They were driving them toward hunters who were waiting behind large boulders. Rows of long poles with flags at their tips blocked the reindeer's potential escape routes. These attachments were meant to *scare* the reindeer, since they are not used to seeing objects moving above the ground in a treeless landscape. But the text didn't answer how the ice patches came into play.

With their thick, double-layered coats and hooves like snowshoes, reindeer thrive in cold, snowy environments, subsisting on lichen during winter and eating willows, grass, and other plants during summer. Biologists have known for decades that reindeer use perennial ice patches as refuge during midsummer, when temperatures are high and biting insects are at their worst. Reindeer are plagued by parasitic warble flies and botflies. From early summer to midsummer, warble flies lay eggs on their coats. When the eggs hatch, the larvae chew through the deer's hides and nestle under their skin, feeding on soft tissues all winter. When they reach maturity the following spring, they emerge from their host to infect another animal. Botflies employ a different strategy, infecting reindeer by laying eggs up their noses. Needless to say, these pests are more than a little annoying; they mercilessly torture the reindeer.

Since insect activity is determined by temperature, insects avoid cold spots and gravitate to warm spots. By spending time on ice patches, reindeer reduce their exposure to these parasites. Even when threatened by predators, reindeer are reluctant to leave these frozen refuges. Pilø concluded that ancient hunters used this habit to their advantage, creating elaborate drive lines with flagged wooden poles placed strategically to frighten reindeer and herd them toward other hunters waiting behind stone blinds with bows and arrows. The arrows were sometimes lost in the snow when a hunter missed his or her shot.

Setting drive lines at ice patches, or anywhere for that matter, must have taken a great deal of effort involving a community of drivers and hunters—but the technique seems to have been extremely effective because Pilø found hundreds of scaring sticks at Juvfonne alone. Of the four hundred artifacts he collected there in the first few years, all but two were scaring sticks.

The Juvfonne ice patch is far from the only ice patch with artifacts in Norway. Pilø has discovered more than four thousand artifacts from sixty-six sites in and around the Jotunheimen Mountains in the years since the discovery of Juvfonne. At some ice patches, Pilø has found entire caches of scaring sticks that had been intentionally left in place, as if they were being stored there until the following hunting season. There's no wood at the elevations where ice patches occur, so hunters would have had to carry the scaring sticks up the mountain. Leaving them there saved time and energy. At other ice patches, the drive lines remain intact and upright, showing Pilø where the hunters had driven reindeer (even though the ice patches have a different shape today than they did hundreds or thousands of years ago).

Unlike the finds at Schnidejoch, just about all ice-patch artifacts found in Norway have to do with hunting—mostly scaring sticks and arrow shafts—with one exception that is reminiscent of the Schnidejoch. From about 300 CE to 1500 CE, people crossed a large patch of ice that blanketed the Lomseggen ridge west of Lom. The ridge connected permanent settlements along the Otta river to summer grazing areas and farms located among the inner alpine valleys. Entire families traveled over this ridge with horses, fodder, and the food they would need before the alpine grass turned green and the farms produced food. The trail is marked by series of stone cairns and has a small stone shelter at the very top in case travelers were caught in a storm. Over the years, Pilø and his colleagues have found snowshoes made for horses, wooden skis, and the remains of horse-drawn hay-sledges.

The Fimbulvetr Felt around the World

Like Hafner, Pilø wanted to know what could account for the rise and fall of hunting artifacts in the mountains. The oldest arrows and scaring sticks were about six thousand years old, and the youngest date to the Middle Ages. But throughout the last six thousand years, there were times when dozens of hunting artifacts disappeared into ice patches and other times when it seemed as if few people had hunted in the mountains. The numbers of scaring sticks and arrows found not just at Juvfonne, but at many of the other surrounding ice patches, were especially high during the fifth and sixth centuries. When Pilø looked at the climate record, he discovered that the rise in hunting artifacts during this time coincided with an extreme cold period called the Late Antique Little Ice Age (536 CE to 660 CE). Like the more recent Little Ice Age, the Late Antique Little Ice Age was triggered by volcanoes.

During the winter of 536 CE to 537 CE, a volcano in Iceland erupted, and it was massive. Scientists think that two other volcanoes erupted in the same year—one in the eastern Aleutians and the other somewhere in the Pacific Northwest. Two years later, another volcano erupted followed by yet another eruption several years after that. The result was a volcanic winter felt around the world. Summer temperatures plummeted and stayed low until 550 CE. In Norway, those years are considered the coldest of the last 2300 years, leading one historian to write that "it was the beginning of one of the worst periods to be alive."

Millions of people in Scandinavia are reported to have died of starvation from widespread crop failures (some historians claim that half the population perished). Roman official Cassiodorus wrote down his observations of the event from Ravenna, Italy, on the coast of the Adriatic Sea. "Something coming at us from the stars" that produces a "blue coloured sun" dims the full moon and results in "a summer without heat ... perpetual frost ... [and] unnatural drought." The crops wither in the fields, nothing grows, and all the while "the rays of the stars have been darkened." The skies over the Mediterranean and Middle East were reportedly so dim that the sun hardly cast a shadow the entire year.

The catastrophe may have even unleashed the Justinianic plague from 541 CE to 549 CE, which affected all of the Mediterranean Basin, Europe, and the Near East. The plague was caused by the same bacterium responsible for the Black Death at the beginning of the Little Ice Age. Plague, famine, and the lack of vitamin D from the dust veil contributed to health problems in populations throughout Europe. The events of the Late Antique Little Ice Age were so severe that this period overshadowed both the Black Death and the 1918 flu pandemic in terms of number of deaths. It was a difficult time to be alive.

The harsh summers following the volcanic eruptions are thought to have inspired the legend of Fimbulvetr, or "awful, mighty winter" in Old Norse. Described in Snorri Sturluson's thirteenth-century book on Norse poetry and mythology, the Fimbulvetr is three consecutive winters with no summer in between, in which "snow shall come from all corners; frosts shall be great then, and winds sharp; there shall be no virtue in the sun." According to legend, Fimbulvetr heralds the beginning of Ragnarok—the end of the world.

So why did Pilø find that more arrows and scaring sticks were lost in the mountains during this horrible period in history? Pilø suspects that failed harvests could have pushed farmers to depend more on hunting wild game in the mountains. Norway is already at the northern limit of agricultural possibility. A plunge into cold summers would have forced people to seek alternative sources of food. Although the Juvfonne ice patch reveals a long history of reindeer hunting, these hunters were also farmers. People had been farming in Norway for the last 6200 years. Pilø has found that the ice patches with the most artifacts are relatively close to settled valleys, usually within a two- to three-hour hike from present-day farms.

But the climate warmed during the Medieval Warm Period, and the number of hunting artifacts found from this period dropped dramatically. Although Pilø isn't sure why hunting at ice patches seemed to decline at this time, there are a couple of possible reasons. One is that Medieval hunters shifted their hunting strategy from ice-patch hunting to harvesting reindeer during their migration. This involved mass trapping of reindeer using pitfall traps and corrals, a technique that may have reduced the need for hunters to visit ice patches. Mass trapping killed far more reindeer than could be used by the local community, and as such, the surplus was exported to meet the demands for reindeer products abroad. High hunting pressure eventually became too much for the reindeer, and their numbers dwindled. By the thirteenth century, the wild reindeer in and around the Jotunheimen Mountains were nearly extinct.

Another possibility is that the warmer climate forced hunters to travel higher into the mountains to harvest reindeer, since there would have been little ice down lower. It's possible that these higher-elevation ice patches, which have yet to melt, contain evidence of reindeer hunting from this period. But Pilø will have to wait to find out until those highest of ice patches begin to melt.

The Resilience of Ice Patches

As Pilø continued collecting artifacts and searching for new ice patches, he began to wonder what makes ice patches tick and whether finding out more about ice patches could help him interpret the archaeological finds. Up to this point no one had bothered much with studying the ice itself. Archaeologists were concerned only with the artifacts preserved in the ice. Nor did glaciologists express much interest in ice patches. Compared to alpine glaciers and polar ice sheets, perennial ice patches just weren't that interesting to them, mostly because no one suspected that the ice was all that old. While Pilø didn't know much about glaciers or ice patches, he knew someone who did.

Rune Ødegård is a cryospheric scientist at the Norwegian University of Science and Technology in Gjøvik. "I don't know much about archaeology, but I know about glaciers," Ødegård told me one spring day over video chat. Plus, he knew the area from working on his master's project during the early 1980s. He had also recently installed a permafrost monitoring station not far from the Juvfonne ice patch.

When the ice tunnel was first excavated in 2010, Ødegård collected samples of reindeer dung embedded in the ice. Because the organic layers were clearly visible, collecting samples from the different layers could shed light on the accumulation of ice over time. He found that the organic layer near the surface (or ceiling) of the tunnel formed about one thousand years ago, and the oldest organic layer dated to 3200 years ago. Based on the age of the oldest organic layer, Ødegård figured that the Juvfonne ice patch probably didn't survive a prolonged warm phase called the Holocene Thermal Maximum, which began about nine thousand years ago and ended roughly five thousand years ago. Temperatures were 2.7°F to 3.6°F (1.5°F to 2.0°C) warmer than today's temperatures, particularly from around 8500 to 8000 years ago. Ødegård told me that pretty much all the glaciers in Norway at this time in history had melted. But when the tunnel was rebuilt (after it collapsed in 2011), Ødegård had a second opportunity to check his initial results.

The ice patch had melted so much that to reach the deepest ice, the former 100-foot-long tunnel became a 230-foot-long tunnel, which meant that Ødegård was sampling dung from deeper inside the ice patch. As before, Ødegård collected organic samples of reindeer dung throughout the column of ice. He figured that maybe the ice would return a date similar to the previous date, but when the results came in, he was shocked. The deepest layer of dung (and by inference the ice below the dung) was 7600 years old, which made it the oldest ice located in mainland Norway to date, and far older than anyone had imagined. "When we drilled Juvfonne we weren't looking for the oldest ice in Norway," said Ødegård. They were just trying to build a picture of how ice patches form and whether the layers they observed in the ice were in chronological order. "It was completely unexpected," he said.

The date of the ice was so surprising because it meant that the Juvfonne ice patch formed at a time when there was very little ice in Norway's mountains. The climate had only just started cooling down after the Holocene Thermal Maximum. How could ice patches form during the latter half of the Holocene Thermal Maximum when glaciers did not? Many of the most productive artifact-bearing ice patches found today occur below the equilibrium line of glaciers—the point at which accumulation equals ablation. Theoretically, perennial patches of ice shouldn't exist below the point at which glaciers melt. And yet there are thousands of them.

Linda Jarrett is a geographer who, as part of her PhD research while at the Norwegian University of Science and Technology, studied the landscape features that influence ice-patch size and growth. She told me that what interests her most is where the natural world meets culture. An ice patch used for hunting is a prime example of the intersection between nature and culture. By the time Jarrett began studying ice patches, the cultural side of things was well-established, but not much was known about the natural part—the ice itself. Jarrett thought she could help. She focused on three ice patches on the Dovre–Oppdal Mountain plateau north of Jotunheimen. These were the same ice patches from which Oddmunn Farbregd had assembled a small collection of artifacts. She set out to understand how ice patches persist at such (relatively) low elevations.

Jarrett applied the same techniques to ice patches that scientists use to study glaciers. One of the tools she used is called a terrestrial laser scanner. Terrestrial laser scanners (also known as LiDAR) acquire three-dimensional coordinates of numerous points on land by emitting laser pulses toward these points and measuring the distance from the device to the target. You've seen them in action if you've ever noticed survey crews using an orange tripod with a UFO-shaped object on top set up along the road, but they have many applications from geomorphology surveys to the fashion industry. In terms of ice patches, the terrestrial laser scanner provided high-precision maps of the amount of snow that accumulated atop an ice patch and how much has melted in a given year.

Jarrett told me that although ice patches accumulate snow through direct snowfall, it's not the primary way they grow. Based on the data collected from the terrestrial laser scanner, Jarrett said that wind was the major reason why ice patches accumulate so much snow.

Mountain and alpine environments are windy. The open terrain, complex topography, and differences in temperature between warmer low-elevation valleys and colder peaks forces wind to speed up as it is driven between saddles and over ridges, especially during winter. Jarrett told me that wind drift alone provided two of the ice patches she studied with at least half of their additional accumulation of snow compared with the surrounding terrain. A single storm event could provide an ice patch with all or most of its annual accumulation of snow.

The accumulation of wind-driven snow on ice patches has to do with local topography. Most perennial ice patches in the northern hemisphere are tucked into hollows on cold north- or northeast-facing slopes. In addition to facing north, they are situated in the lee of prevailing winds, which means that winter and spring winds pick up and deposit most of the snow that accumulates on an ice patch. "Without wind drift, there are no ice patches," Jarrett told me.

During her study, Jarrett found that the ice patches accumulated between twelve and eighteen feet of snow during a single season. In some parts of the ice patches, an astonishing twenty-six feet of snow was amassed. It's this wind-driven accumulation that allows perennial ice patches to persist below the equilibrium line of glaciers.

Pilø told me about extreme winds he experienced on one late-season visit to Juvfonne in 2009. The winds gusted so strongly he had trouble staying upright. Undeterred, Pilø continued his survey but was forced to lie atop the boxes containing artifacts when the wind gusted. A weather station near the ice patch recorded wind speeds of up to 145 miles per hour (65 meters per second). The wind had been so extreme that rocks picked up by the wind smashed windows in the nearby lodge. And because the weather station itself snapped (and stopped collecting data), it may have been even windier.

But despite wind drift, ice patches don't grow indefinitely. If they did, they might become glaciers. If these ice patches had been glaciers at some time in the past, then the artifacts would be smashed to pieces, but they aren't. They look as though they were dropped moments ago. Jarrett found that the same factors that allow ice patches to persist also control their growth. In other words, ice patches have a maximum accumulation potential. Any excess snow atop an ice patch that is not protected by topographical features will blow away.

Imagine a perennial ice patch during a high-melt year. Melting ice would leave a depression on top of the ice patch—a hollow where the snow from previous winters had been. Although there is less snow and ice, it doesn't melt completely because, said Jarrett, "the ice patch pulls back into a shadow provided by local topography." In other words, the landscape immediately surrounding the ice patch provides just enough shade to prevent it from melting completely. The following winter there is a big hole to fill. Even in a low-snow year, an ice patch can still accumulate a lot of snow due to wind alone. So, ice patches are resilient to changes in climate over the long term.

Some scientists view perennial ice patches as existing on a continuum between ephemeral snow patches and glaciers (in some circles, ice patches are known as "glacierettes"). Jarrett says it's not that simple. While it's possible that ice patches can become glaciers and glaciers can become ice patches, the most productive ice patches—those with the most artifacts—were probably always ice patches. Otherwise, they would have destroyed the objects within them. From the lichenometry of the ice patches Jarrett studied, she found that while the ice patches were considerably larger during the Little Ice Age, they weren't large enough to overcome topography. They had always been ice patches.

Jarrett observed that ice patches seem to require a larger shift in the regional climate, from one climatic envelope to another, to disappear completely or grow into glaciers. Glaciers are so large that they possess an inherent buffer against annual fluctuations in temperature and precipitation. And unless the climate is consistently warm or consistently cold for a decade or longer, you won't see dramatic changes to the glacier's size or length. But with ice patches, it's the opposite. Annual fluctuations cause huge changes to ice patches. They respond instantly to years with a lot of snow, especially if it's windy. Jarrett calls ice patches "weather-sensitive but climate-resilient," whereas glaciers are "weather-resistant but climate-sensitive."

But Ødegård tells me there is another factor in addition to wind that explains why ice patches survive below the equilibrium line of glaciers: permafrost. Permafrost, which is essentially frozen soil, lies beneath roughly a quarter of Norway. In alpine environments, permafrost is patchy because of complex topography and rapid shifts in elevation. Ødegård says that long-lived ice patches must be cold at their bases to withstand dramatic changes in climate over thousands of years. So, the permafrost insulates the ice patch from below, which also helps retain the permafrost, and the annual accumulation of snow helps insulate the ice patch from above.

Ødegård says that in winter, the horizon between the ground and the ice patch will dip a bit below zero, and during summer that interface can't rise above zero—otherwise the ice patch would melt. He suspects that the permafrost provided the conditions for ice-patch formation and the wind provides the mechanism for snow accumulation. "If you look at the limits of where permafrost is today, all long-lived ice patches are at that threshold or well above it," says Ødegård.

Although ice patches and glaciers are distinctly unique, ice patches do move, at least incrementally. "We were doing field work one day and then we saw scratch marks on the ground where the ice patch had receded," Jarrett told me. Even though there were no moraines at the three ice patches Jarrett studied (the presence of a moraine would indicate that an ice patch had once been a glacier), there were abrasions on the rock that had up until recently been covered by the ice patch. "We were really surprised at how much the ice patch slid across the rock," said Jarrett.

As I walked the halls of the Juvfonne ice tunnel, I could see the tilted layers of ice toward the bottom of the ice patch. All throughout the walls were bands of dirty ice where dust, dirt, and dung had accumulated over millennia. As I headed toward the exit, I passed under the Jörmungandr and wondered, How long before it releases its tail and the tunnel collapses? How many years until the ice patch is too thin to support a tunnel?

Canary in a Coal Mine

I stepped out of the Mimisbrunnr ice tunnel and into the bright light of day. The sky had cleared, and the sun glared off the ice patch. I pulled my sunglasses down over my eyes. As I looked back at the ice patch and the snow-covered mountains beyond, it was hard to imagine this landscape of giants without snow ... and yet we are in the age of melt. A single year with huge amounts of snow or scant snowfall makes little difference in the long run. That's just the normal variability of the environment. But, when there are many years that tend toward one direction or another, even with variability between years, we know that things are changing.

Since 1900, Norway's average temperature has increased 2°F (1°C). While precipitation has increased 20 percent since 1900, more of it is falling as rain than snow. From 2009 to 2014, the Juvfonne ice patch lost more than 30 feet of depth and 170 feet in length. For seven out of the ten years between 2010 and 2020, the mass balance (the amount of snow gained minus the snow lost in a given year) of the Juvfonne ice patch has been negative. If this continues, the 7600-year-old ice patch, and tunnel, will disappear just like the one at Schnidejoch pass.

Pilø told me that when he first got into ice-patch archaeology, he wasn't much into the climate thing. "It was just 'OK, the ice is melting. Let's go and find the artifacts.'" But he says that the climate aspect has become a much bigger part of his work since then. "Ice-patch archaeology is a canary in the coal mine, showing that there is something happening in the high mountains, which is quite disturbing," said Pilø. "One question I have is how are the ice patches going to go? Slowly? With a bang? Let's say we get three or four years of very hot summers and little winter precipitation. Will that be enough to kill them off? I don't think anybody has the answer to that."

Norway's ice patches have revealed a story of reindeer-hunting over the last six thousand years. Today, hunting reindeer is tightly controlled by the Norwegian government. Centuries of overhunting, domestication, development, and now climate change have reduced the wild population to about twenty-five thousand, all of which roam the mountains of southern Norway.

Most of Norway's reindeer are herded by the Indigenous Sami in far northern Norway. Although the majority of the Sami live in the north today, they also historically lived in southern Scandinavia until around the Middle Ages, when the Germanic tribes pushed them north. Archaeologists have found artifacts of probable Sami origin in southern Norway that suggest they lived there at least until one thousand years ago. It's possible that both Norse and Sami people lived in the mountainous areas of Southern Norway at the same time, which makes attributing artifacts to Sami or the Germanic tribes difficult.

Sami reindeer-herders move with their animals in annual cycles between summer and winter pastures. They transitioned from hunting wild reindeer to living as semi-nomadic pastoralists sometime during the sixteenth or seventeenth century. Some anthropologists think that this shift occurred in response to a combination of high taxes; growing demand for reindeer products like pelts, combs, and antlers; and mass trapping, which led to a declining wild population. Herding provides not only for the herders' personal needs but is also a form of wealth to pay taxes. However, living traditionally is getting harder.

In recent years, Sami herders have noticed dramatic changes in climate, especially during winter. Warmer winters means greater snowmelt. Meltwater then re-freezes into an impenetrable layer of ice when temperatures inevitably drop again. Reindeer have a hard time breaking through this layer of ice to forage for lichen, and many starve if not given supplemental food.

Halfway around the world, a different but just as close relationship with reindeer (known as caribou in North America) has been reinvigorated through ice-patch archaeology, with artifacts dating close to the last ice age. For the First Nations of the southern Yukon, ice-patch artifacts are heirlooms that connect them to their ancestors.

SIX

Yukon Hunters of the Alpine

It was late summer 1997, and Kristin Benedek and her husband Gerry Kuzyk were hunting Dall sheep in the Southern Lakes region of Canada's Yukon Territory (now known simply as "the Yukon"). The Southern Lakes region lies in the rain shadow of the Saint Elias Mountains. It's dry and windswept, a landscape of rolling hills, pine forests, and long, narrow lakes. In spring, mountain snowmelt rushes into rivers that thread their way through valleys of willow and aspen. In the highest elevations, well above treeline, Dall sheep graze on what few plants grow in this rugged and vast landscape. Benedek and Kuzyk followed their trails, hiking many miles to find them in this remote roadless area.

They had spotted several rams on the slopes of Thandlät Dhal, which means "sharp-pointed mountain" in Dákwanjè (Southern Tutchone)—a language spoken by the Champagne and Aishihik First Nations, whose traditional lands they were hunting. As they drew near, they noticed a thick black rind along the base of a large patch of ice. By summer's end, patches of perennial ice and snow are often coated in a layer of windblown dust, but the ice on Thandlät Ddhal was black rather than the color of dirt, and it smelled like a barnyard.

Kuzyk, a wildlife biologist with the Yukon government, was intrigued. It looked to him like caribou dung (and lots of it), but he

118

knew that caribou hadn't been seen there for more than sixty years, and the nearest herd lived at least thirty miles to the southeast. Yet this dung looked fresh. There were even whole pellets that appeared as though they'd been dropped moments before.

Caribou—*mäzi* or *udzi* in Dákwanjè—is a French word believed to originate from the Mi'kmaq word *Xalibu*, meaning "shoveler or pawer of snow." During winter, caribou use their wide hooves to paw away snow and graze on the lichen beneath. Some caribou populations migrate to find food, but others, like those in the Southern Lakes, find adequate food by moving short distances between high-elevation summer grounds and low-elevation wintering areas.

Caribou have been absent from the area for sixty years, but First Nations Elders remember a time when they were so numerous that it looked as though the whole mountain were in motion. Jimmy G. Smith, Champagne and Aishihik First Nations (CAFN) Elder, saw some of the last caribou in the area in the winter of 1932. CAFN Elder Mrs. Annie Ned confirmed Smith's account. In 1985 she told anthropologist Julie Cruikshank that "there used to be lots of caribou, even in my time [1930s]. When caribou came, it was just like horses. You could hear [their hooves] making noise on the ice."

Yukon's Southern Lakes is a culturally diverse landscape, home to six First Nations of Tlingit and Athabascan ancestry. Before outsiders came to the southern Yukon, Tlingit and Athabascan people lived a subsistence lifestyle that involved fishing, hunting, and gathering. Home was a seasonal round of villages or shorter-term campsites situated along lake shores and the mouths of salmon-spawning rivers. Mobility was essential for Indigenous peoples to access seasonally abundant foods. Elders often referred to travel as a fundamental part of their lives. "People used to walk around all the time," according to Mrs. Annie Ned. Caribou were an important part of that subsistence lifestyle. They still are.

Caribou populations (and their genetics) have been shaped by ice-age cycles. At the start of the last glacial cycle one hundred thousand years ago, caribou thrived, with the total population numbering in the millions. As ice sheets expanded, the caribou's range contracted. During ice-age cycles, caribou survived in isolated, ice-free refugia. But when the ice sheets shrank, the caribou's range grew, and the various populations were reunited. Some populations bred with one another, while others remained separate. Eventually, this expansion and contraction over multiple ice-age cycles led to two distinct types of caribou: barren-ground caribou and woodland caribou. The Southern Lakes area is one of many regions that is home to a type of woodland caribou, although barren-ground caribou may have also lived there at one time.

By the early 1990s, though, there were only about four hundred caribou left among three Southern Lakes herds. Decades of overhunting, habitat loss, barriers to migration (such as fences and roads), and fire suppression led to their near extirpation from the Southern Lakes area. The decline in caribou began with the arrival of forty thousand get-rich-quick hopefuls who descended on the Yukon during the Klondike gold rush of the late 1890s. As gold-seekers, fur trappers, hunters, and missionaries came to the Yukon, others followed close behind. They built roads, churches, and houses to accommodate the growing settler population. Local Indigenous hunters sold caribou and moose meat to the newcomers. From the 1910s to the 1940s, big-game and trophy hunters arrived to partake in the region's seemingly endless supply of wildlife. And then came construction of the Alaska Highway, which brought even more people to the area when the road opened to the public in 1948. The network of roads not only provided hunters with easy access to caribou, but the roads also led to more frequent vehicular collisions with caribou. The population of Whitehorse jumped from 750 residents to twenty thousand residents. Over the next several decades, these pressures contributed to declines in caribou.

"In the 1970s and 1980s there was a lot of mining, and they were putting in a lot of roads. After that, people had more access to the mountains with quads and ATVs. They just opened the country right up," said hunter, trapper, and Tlingit master carver Keith Wolfe Smarch when I phoned him one winter morning. He spoke to me from his workshop in Whitehorse over a crackling landline. "We used to get massive herds right here across from my studio, but they don't come here anymore."

The situation became so dire by the early 1990s that the governments of the six First Nations whose traditional lands overlap the three Southern Lakes herds voluntarily stopped hunting caribou, and the Yukon government issued a closure for licensed hunters. When Benedek and Kuzyk found the ice patch covered in caribou dung, it had only been five years since the hunting ban went into effect, and caribou numbers were at an all-time low. When Kuzyk returned to work in Whitehorse a few days later, he sought the help of caribou specialist Don Russell, who worked for the Canadian Wildlife Service at the time. Kuzyk's supervisor with the Yukon government, Richard Farnell, was out of town, and Kuzyk felt his discovery was too important to wait. Russell agreed. Two weeks later they flew by helicopter out to the ice patch, and Russell knew right away it was caribou dung. "Sure enough, it smelled like a barnyard," Russell told me one day over video chat.

Russell, who has since retired, boasts a head of thick, curly white hair and a full beard to match. As Russell considered the enormous volume of caribou dung, he noticed a twig poking out of the snow. "What's this?" he asked as he bent down and held up the broken stick. It was about four inches long with a bit of string twisted around one end. Russell thought it might be part of an arrow shaft, but neither Kuzyk nor Russell were sure, so they collected the wood along with a caribou jawbone, an unidentified bone fragment, a clump of hair, and some dung pellets.

At the end of September, Russell and Kuzyk returned to the ice patch with glaciologist Erik Blake, whose company (Icefield Instruments Inc., based in Whitehorse) develops specialized instruments to drill cores in polar ice sheets. Kuzyk and Russell wanted to know if the caribou dung only appeared on the surface of the ice or if it was dispersed throughout. Using a hand-cranked auger, Blake collected several cores from the ice patch. The cores showed layer upon layer of caribou dung interspersed with clean layers of ice.

Russell knew that caribou congregated on ice patches during late summer to cool off and escape mosquitoes and warble flies, but the explanation for such dense quantities of dung accumulating and persisting in a place where caribou had been absent for decades was a mystery. He reasoned that the dung must be at least sixty-five years old, since that's the last time caribou had been seen in the area, but given the depth of the ice and dung, he wondered if it could be much older. And then there was the piece of wood, which was obviously handcrafted. Kuzyk knew there was a story there, so he took the piece of wood to Greg Hare, an archaeologist with the Yukon government in Whitehorse.

"Gerry [Kuzyk] came into my office a day or two later and said he made this discovery and wanted me to have a look at it," Hare told me over video chat one afternoon in late autumn. He immediately recognized the exceptional craftsmanship of the object, which had been found in three pieces that fit together like a puzzle. On one of the pieces of wood, two or three strands of sinew had been affixed to the shaft in a Z twist. But what most intrigued Hare was that the sinew still held two bird feathers to the shaft. While only the quills remained, it was the most well-preserved artifact he'd ever seen.

Judging from its extraordinary preservation, Hare thought that the artifact was probably the shaft of a First Nations hunting arrow from around the time of the gold rush. "I figured it might be a hundred years old at most because we all thought that the ice wasn't that old," said Hare. A 100-year-old shaft would make it unequivocally part of an arrow. Hare told me that the bow-and-arrow is a hallmark of the Late Precontact period. To find out just how old it was, Hare sent a small chip from the suspected arrow shaft along with fecal pellets from the bottom of the ice core to IsoTrace Laboratory at the University of Toronto for radiocarbon dating.

Tools of the Trade

Prior to European colonization and the introduction of guns, the bow-and-arrow was the most recent technological innovation adopted by Indigenous peoples in the western hemisphere. But First Nations arrived with a tool called an atlatl, also known as a spear thrower. Although the word atlatl comes from the Nahuatl language of the Aztecs, the tools were used by ancient hunters all over the world (although none have been recovered from Africa, at least not yet). No one knows just where the first atlatl arose, but the earliest example was found in a French cave called Isturitz. That atlatl was made 17,500 years ago from the antlers of an adult male reindeer.

An atlatl consists of a dart that is launched using a throwing board. The throwing board is made of a straight piece of wood, roughly two feet long with a handgrip at one end and a spur at the other end. The spur is a point that fits into a cavity at the back of a four- to six-foot-long dart. The dart is suspended parallel to the board and is held by the tips of the fingers at the handgrip. A hunter then launches the dart with a sweeping arm and wrist motion similar to a tennis serve.

The addition of a throwing board was an important modification to the hand-thrown spear because it increases the leverage of the thrower's arm, propelling the dart faster and farther than body force alone. A skilled hunter could make a successful kill from up to about sixty feet away with an atlatl. Although atlatls were better than arrows at puncturing an animal, when the bow-and-arrow was introduced, those who encountered it usually chose the new weapon because it offered several advantages over the atlatl.

For one, the bow-and-arrow is an overall smaller hunting tool than the atlatl. Downsizing to the bow-and-arrow meant that a hunter could carry a dozen or more arrows versus the two or three darts used with an atlatl. Another advantage is that the bow-and-arrow's smaller design allowed hunters to shoot from a kneeling position and conceal themselves behind a shrub or stone blind (although First Nations hunters along the Pacific coast used similar weapons to hunt marine mammals from a sitting position in a boat). Lastly, an arrow shot with a bow can travel up to a hundred feet, which means a hunter doesn't have to get as close to their target.

Atlatl or Arrow?

Two months later, Hare received the results of the radiocarbon analysis. The fecal pellet dated to 2535 years before the present, and the wooden shaft dated to 4953 years before the present, give or take fifty years. "All of a sudden we realized that we had uncovered a phenomenon of not just organic preservation, but cultural preservation that went back thousands of years," said Hare.

Given the date of the wood Russell had picked up at the ice patch, Hare suspected that the tool was part of an atlatl and not an arrow shaft as originally believed, but at the time the possibility of the greater antiquity of the bow-and-arrow was an open question for the Southern Lakes area. Hare said that the introduction and onset of bow-and-arrow technology in the new world has always been a very popular and controversial topic for archaeologists.

Points, and whether they are made of stone or bone, are the standard archaeologists use to determine tool type and the relative age of an object. In general, shorter and lighter arrows require lighter and smaller points, and heavier and longer darts require heavier and larger points. For this reason, archaeologists often attribute larger points to atlatls and smaller points to arrows. The problem is that there is overlap between the size of an atlatl point and the size of an arrow point or arrowhead.

Since archaeologists usually have only the stone or bone points to go on, there is a lot of room for different interpretations. While bone can be dated, if archaeologists don't know when First Nations made the transition from one tool type to another and an artifact is missing the entire rest of the weapon due to decay over time (which is almost always the case), then how does one know if the point was hafted to an arrow shaft or to an atlatl dart shaft? This one tool couldn't answer that question—at least not yet.

* * *

By 1997, Ötzi the ice mummy was well-known throughout the world, but not much had melted out of alpine ice since—at least nothing that Hare knew about. He wondered whether the discovery of Thandlät was a one-off find like Ötzi seemed to be or if it was a harbinger of more to come.

The following year, Hare flew with pilot Delmar Washington, a CAFN citizen, to help him survey the Southern Lakes for other promising ice patches. The region he planned to survey was hundreds of square miles of remote wilderness salted with hundreds if not thousands of perennial ice patches. Thandlät was first on the list of sites to visit. It was July 23, and the weather was good for flying. Washington piloted several people out to the ice patch, including CAFN Elder Jimmy G. Smith, CAFN heritage officer Diane Strand, and biologists and archaeologists.

"There were a lot of people out there that day," Strand told me. During the first years of the Yukon ice-patch project, Strand served as CAFN's heritage officer. Between 2006 and 2010, she held the office of Chief for her First Nation. She's now the owner of Crow's Light Healing and Discovery, where she offers therapy that integrates spiritual, emotional, and physical healing. She spoke to me over video chat from her home in Haines Junction.

In 1998, the CAFN was a newly formed government. Just five years earlier they had settled comprehensive land claims and self-governing agreements with the Yukon government and the government of Canada.

These agreements, which are modern-day treaties, went into effect in 1995.

"I just remember looking around and seeing all these non-native people there," she went on. "And we had taken Elder Jimmy G. Smith with us, and he told a story about caribou and how many he remembered being there. And one of the biologists said 'Well, that's not true, because there were no caribou at that time.'" Strand's defenses went up at that moment, so she hung back from the group.

Strand told me that Indigenous peoples' histories are constantly forced to fit into a colonial structure rather than archaeologists using an Indigenous framework. An Indigenous framework includes more than just study of the material objects. The knowledge of how these objects were made, oral traditions, performing arts, social practices, rituals, and worldviews about nature and the human place in it are all important aspects of an Indigenous knowledge system. And the artifacts melting out of the ice patches are just one part of their knowledge framework.

As Strand walked the edge of the ice patch, she sought help from her ancestors. "I asked for guidance because I felt like there was a violation of visitors onto this land and I didn't want to blow up." Although Strand and Hare had a good relationship, tensions between CAFN and the Yukon government at the time were high. One major point of tension was disagreement about what constituted part of CAFN's (and other Yukon First Nations') heritage. If an artifact is ten thousand years old, for example, is it of Tlingit or of local Athabascan origin, or could it belong to a group that migrated far to the south like the Navajo?

The Yukon government and the government of Canada viewed ancient artifacts as a shared cultural and scientific resource, but Strand and her colleagues didn't see it that way. "In my mind, what I had always told any of the students that had worked with us, is that your ancestors made this. And even if it was Navajo, we are related to them, we are all of the same Athabascan family group."

The Yukon ice-patch project began a year after the discovery of Kennewick Man, also called the Ancient One, on the banks of the Columbia

River in the state of Washington. The Umatilla people upon whose traditional lands Kennewick Man was found, along with several other local tribes, wanted to bury him according to their customs. But the United States government scientists were eager to study him because his remains were far older than most human remains found in the Americas at that time. The Ancient One dated to 8700 years before the present.

In 1990, President George H. W. Bush signed into law the Native American Graves Protection and Repatriation Act (NAGPRA). NAGPRA stipulates that the tribes are entitled to the return of any remains stored or on display in museums, universities, and other institutions. Descendants and affiliated tribes can request the return of ancestral remains for proper burial. Hundreds of sets of human remains were repatriated following NAGPRA (although many still remain at institutions despite requests from tribes). While most of these remains were relatively recent—no older than a few centuries or maybe a few thousand years at most—scientists claimed that no tribe could reasonably claim ancestry to Kennewick Man since the remains were so old.

Based on an initial examination of the Ancient One's morphology, scientists concluded that he most closely resembled people of Polynesian or Southeast Asian descent, which is as absurd as it sounds. Ever since Europeans colonized the western hemisphere, archaeologists and anthropologists have doubted the ancient existence of Indigenous peoples in what are now the Americas, despite substantial evidence to the contrary (including, most recently, 22,000-year-old fossilized human footprints discovered at White Sands National Park in New Mexico). If the Ancient One was not closely related to present-day Indigenous peoples of the Americas, it meant that NAGPRA wouldn't apply, and scientists would be free to study the remains without restrictions.

It took more than ten years in court and advances in extracting ancient DNA to settle the argument. The results were unequivocal. The Ancient One was related to present-day First Nations. Although his remains were eventually returned to the Umatilla people, the idea that First Nations people only recently arrived in the western hemisphere hasn't faded—despite dozens of exceptionally old archaeological sites found throughout the western hemisphere.

As Strand hung back from the group, she refocused her attention on the ground where the ice gave way to rock. "I was looking down and saw something that looked a little bit different. I picked it up and brushed the mud away and said, 'I found something.'" Strand had found the second ice-patch artifact—a dart fragment with sinew and ocher bands staining the wood. Radiocarbon sampling of the artifact returned a date of 7705 years before the present. She said that all those "ologists" walking ahead of her had missed it, but she hadn't. Every time there was a significant find during her time with the CAFN government, Strand said that it was an Indigenous person who made the discovery. "So, to me it was meant to be, like our ancestors were looking out for us."

A Process of Remembering

As Hare continued mapping perennial ice patches during his helicopter surveys, he and his colleagues with the Yukon ice-patch project noticed that most of the patches were tucked into hollows in north- or northeast-facing slopes. In the northern hemisphere, north and northeast faces receive less sunlight than south- and west-facing slopes do, which helps conserve ice. Hare also noticed that ice patches tended to occur in cirques—amphitheater-shaped basins carved by glaciers. A classic cirques is usually high on a mountainside and surrounded by steep cliffs on at least three sides. The result is that ice patches in cirques facing north or northeast stay frozen throughout the summer, just like Jarrett had discovered in Norway.

Hare knew that not every ice patch would yield artifacts, so he started with what he knew about the ice patch that did yield artifacts: Thandlät. He reasoned that if people had been hunting caribou that were congregating on the ice—which would explain the caribou dung and the wooden shaft being found together—then perhaps other dung-rich patches might yield artifacts. He was right. Hare found that the abundance of artifacts was strongly related to how much organic matter was visible on the ice. The more caribou dung, the greater the probability that the ice patch held artifacts. In other words, more caribou meant more hunting opportunities.

But while caribou congregate on many different types of ice patches, not all ice patches are good grounds for hunting. The best ice patches are those that lie horizontally across a slope, just below a ridge. The natural topography of these so-called "transverse ice patches" allows hunters to approach from the back side, where they can stay hidden. Furthermore, prevailing winds carry scents up the mountain in the afternoon, so hunters would be upwind of the caribou and largely out of sight if they approached at the right time of day.

During the first two years of surveys, Hare found thirteen artifact-bearing ice patches yielding more than a hundred organic artifacts. "This wasn't just a flash in the pan," said Hare. "This was widespread." But there wasn't a lot of time to examine the finds. Hare was too busy collecting artifacts and figuring out how to store these objects so they wouldn't degrade. He saw his role during those early years as the facilitator of a rescue mission. "Our job ... was to get out there to recover as many artifacts and identify as many ice patches as we could, because who knows what's going to happen next year?"

Hare told me that when artifacts began melting out of Yukon's ice patches in 1997, climate change and melting glaciers were at the forefront of scientific and political discussions. The IPCC published their first assessment report in 1990, stating that "emissions resulting from human activities are substantially increasing the atmospheric concentrations of ... greenhouse gases ... resulting on average in an additional warming of the Earth's surface." And in 1992, more than 1700 scientists signed the World Scientists' Warning to Humanity issued by the Union of Concerned Scientists, which opened with the words "human beings and the natural world are on a collision course."

Those initial thirteen ice patches marked the beginning of the Yukon ice-patch project, which has been a collaboration between Yukon government archaeologists and First Nations governments from its inception. Although that first ice-patch artifact was found on CAFN's traditional lands, within the first several years of the project, artifact-bearing ice patches were found in the traditional lands of five other First Nations, including the Carcross/Tagish First Nation, the Kwanlin Dün First Nation, the Kluane First Nation, the Ta'an Kwäch'än Council, and the Teslin Tlingit Council.

Similar to Norway's ice patches (with a few exceptions), the common thread among Yukon's ice-patch artifacts is that they are all related to big-game hunting. When the ice-patch project began, Strand and other First Nations heritage officials scoured their archives for references related to hunting caribou specifically at ice patches. They also interviewed Elders to see what they recalled or remembered hearing from their parents and grandparents about caribou hunting. Although caribou were harvested year-round (using snares to capture individual animals or hunting-fence complexes that funneled migrating caribou into a trap where they could be easily harvested), only one anecdote surfaced in which a hunter describes harvesting caribou on an ice patch in the 1800s. Strand isn't sure why they found only that one reference, but it could be that land-use patterns changed after caribou populations declined, and so ice-patch hunting dropped out of the oral history. Or maybe hunting on ice patches was not considered all that different from other types of hunting. Regardless, Strand writes that the ice-patch discoveries "triggered a process of remembering and bringing forward the relationship between caribou and people."

Lawrence Joe, director of the department of Heritage, Lands and Resources for CAFN at the time of the ice-patch project, told me that when they'd bring these arrows and darts back to his community, all the Elders wanted to touch them. "It's the kind of thing that makes an archaeologist's hair stand on end," he laughed. The Elders wanted to experience the object, not just look at it from a distance. While some of the artifacts from the ice patches are at least as old as the Ancient One, many are also from the relatively recent past. Joe recalled one of the oldest Elders saying, "It's just like my grandpappy used to make," while handling an arrow shaft from an ice patch. Perhaps some of the Elders *were* handling arrow shafts that their grandfather or great grandfather or great, great grandfather had made. These objects are much more than material artifacts; they are cultural heirlooms.

On the one hand, ice-patch archaeology's connection to ice patches and organic objects has helped spur a cultural revival among First Nations. But there is also important knowledge that comes from speaking to First Nations Elders about these objects: Archaeologists gain a deeper understanding and appreciation of what these items represent, and more practically, how some of the more "obscure" objects were used.

For example, one day, ice-patch surveyors in the Northwest Territories came across a piece of sinew strung in a loop. They were unsure of its use, but it had obviously been made by human hands. That day, First Nations Elders were teaching traditional hunting techniques to a group of Indigenous students at a youth camp. When the ice-patch archaeologists arrived at the camp to ask about the looped piece of sinew, they were surprised to find a modern version of the same tool, which the Elders were using to ensnare ground squirrels. They concluded that the looped piece of sinew they found had once been used as a ground-squirrel snare.

This story illustrates how traditional knowledge and scientific disciplines can complement one another to provide a more complete and culturally sensitive understanding of culture and the relationship of cultures to the land. Strand told me that as much as the ice-patch work helped bring important cultural material and knowledge back to her Nation, the archaeology community also learned to be more receptive to First Nations and their role in archaeology.

During the early years of the ice-patch program, Joe attended science camps for First Nations youth hosted by CAFN. "We brought kids from all over Canada up to the mountains, we flew them up to Thandlät ice patch and hiked them down off the mountain and explained some of the history of the area of the artifacts and how to travel across the land," Joe told me. He taught them how to make an atlatl, and they'd have contests to see who could use them to throw spears the best. "Some of the kids were pretty amazing at how skilled they became with their atlatl. It's old technology but it's still relatable to them."

When I spoke with Strand and Joe, it was obvious that these artifacts don't simply represent history or things of the past. They represent ancestors, memories, and relationships with the land that transcend time. The ephemeral nature of these hunting tools makes them that much more special. These tools had been preserved for many thousands of years by the ice, and they retain not only their shape and function but also the artistry and skill that their maker possessed. Many are decorated with ocher staining, colorful bird-feather fletching, and carvings in the wood. More than artifacts, these objects are threads that link the past to the present while providing a tangible record for future generations of ancestral use of the land.

Another Ice Mummy

Toward the end of the second field season, Hare received a phone call from a hunter by the name of William Hanlon. It was a Saturday evening in August 1999. Hanlon and two companions had been hunting Dall sheep in Tatshenshini-Alsek Provincial Park in British Columbia. The park lies just south of the Yukon border, in the traditional lands of the Champagne and Aishihik First Nations. At 2.4 million acres, Tatshenshini-Alsek Park feels boundless with its glacier-covered peaks, frigid rivers, forests of Sitka spruce, and vast stretches of alpine tundra. Silvertip grizzly bears wander open meadows. Mountain goats defy gravity and slate-backed peregrine falcons tip their wings and wail for their mates as they soar across cerulean skies. Few modern trails score the land, save for one six-mile-long trail to an overlook of the Samuel Glacier.

Hare listened as Hanlon told him about an extraordinary find near a glacier deep in the wilderness. Nine days earlier, on August 14, Hanlon, Mike Roch, and Warren Ward had waded into the swollen Tatshenshini River. They each shouldered a heavy pack filled with camping gear, hunting equipment, and a ten-day supply of freeze-dried food for what they called "the hunt of a lifetime." For six years they had applied for a group tag for the chance to harvest Dall sheep in the park. With tags in hand, they hiked nine miles on their first day through some of the most unspoiled terrain in North America. They even managed to find sheep before sundown.

On the fifth day, they spotted a group of rams high up in a hanging basin at the headwaters of a glacial stream that flows into the O'Connor River. They hatched a plan to go after the rams the following day. Two of the hunters had successfully harvested a ram and were looking to fill their third tag. The hike to the basin would take them most of the day. Because they wouldn't arrive until afternoon, their plan was to approach the sheep from above rather than from below. In the afternoon, warm mountain air drifts upslope. If they approached from below, the wind would have carried their scent straight to the sheep, and the hunters would never see them again.

The trio crossed into a rocky valley strewn with ice patches. Calving ice, mud, and the rocky terrain made travel across the edge of the ice treacherous. Their boots sank into the newly ice-free ground. The summer of 1999 was the warmest summer ever recorded in British Columbia—a record that has since been broken many times. The hunters stopped briefly to refill their water bottles and climbed toward the ridge. At the top of the ridge, they gained their first distant views of the basin. The basin's centerpiece is the Samuel Glacier. The Samuel Glacier is the source of three major rivers: the Tatshenshini, Parton, and O'Connor rivers.

While snacking on beef jerky and trail mix, Hanlon suggested that they traverse an unnamed glacier to cut the distance between them and their destination in the basin. The glacier sat high on a col more than 6400 feet above sea level; it flowed north before making a sharp turn to the west, where it continued a short distance down the valley before terminating. Roch and Ward refused to take the shortcut. They had promised their wives that they wouldn't take any unnecessary risks, and that included crossing glaciers with potentially deadly crevasses. Hanlon relented, and they instead walked along the edge of the ice, which towered above them in a thirty-three-foot wall of turquoise undercut by a deep meltwater channel.

As they walked along the edge of the ice, Hanlon spotted a stick. The stick was notable because they were well above the treeline, so it must have been brought there. He thought it might be a walking stick from another hunter. A few minutes later Roch found a matching stick. They put the two halves together and they fit perfectly. A few feet ahead, Roch found yet another stick. This one was about three feet long and curved, blackened on one end and carved on the other. Roch grew quiet. He looked up at his companions and said, "These aren't sticks; they're artifacts."

Roch thought that the curved stick might be part of an atlatl. Meanwhile, Ward scanned the ice patch with a pair of binoculars. "I think I just found the poor fellow who lost all this stuff." Because Roch and Ward had promised their wives that they wouldn't cross a glacier, Hanlon found a chute in the ice and climbed up to the body. He stood above what looked like the hide of an animal. It was just a brown mass, all crumpled and wet. It was impossible to tell what it was. Nearby lay a small, wood-handled knife. Hanlon picked it up and slid it out of its sheath. It had a leather string wrapped round the handle and a short, corroded iron tip at the other end.

As Hanlon called out descriptions of what he was seeing, the other two men couldn't resist and clambered up to meet him—their promises to their wives forgotten. A human pelvic bone protruded from the ice and was attached to a pair of legs, which were still frozen in the ice. It was a scene reminiscent of Ötzi's discovery. The men put a few of the loose items into a plastic bag, including the knife, to prove to archaeologists that they had found something significant. Although they trembled with excitement from discovering human remains, they tried to refocus their efforts on hunting. They climbed down the ice and began looking for rams with binoculars, but by then the rams had disappeared, so they began a long trek back to basecamp.

Two days and twenty-five miles later, the hunters were back at their truck wondering where to drop the artifacts. Ward found a brochure for the Yukon Beringia Interpretive Centre on the backseat. One of them probably picked it up at a rest stop or a gas station as they passed through Whitehorse at the start of their trip. The center seemed like the right place to report their discovery. Plus, they had to go through White-horse on their way home anyway. It was eight o'clock in the evening when they arrived, and the center was closed, but an intern vacuuming the floors let them in and gave them Hare's phone number.

After hearing Hanlon's story, Hare offered to meet them the following morning. When they arrived at the center, Hanlon showed Hare the items they had collected from the ice patch, including the wood-handled knife. "It was the most active time of the ice-patch project. We were flying every day out to different patches, and we were coming back with dozens of artifacts. We just couldn't believe how widespread the phenomenon was. And all of a sudden there's this guy coming in saying he's found a bunch of artifacts associated with a human body," said Hare.

Finding human remains in an archaeological context is exceedingly rare, and finding well-preserved human remains is even rarer. "I mean, I suppose at one level, I should have believed anything at that point," Hare told me. Still, he was skeptical. It was much more likely that this person's death was recent. He first needed to confirm the hunters' reports, and for that he'd need to see the site and the remains for himself. He set to work organizing a visit to the site by calling Strand.

"Hare phoned me, and he asked me if I was sitting down. I was like, 'Yeah,' and I was making my bed at the time. So, I sat on my bed. I'll never forget, I was looking out the window when he told me that these hunters had come into the Beringia Centre that day, and what they had found. I don't even know if I made my bed after that, because then I was immediately on the phone with members of our government," Strand told me.

On August 17, just three days after the discovery, Delmar Washington flew seven people to the northeast corner of British Columbia. It was a rainy day with low cloud cover—not ideal weather for flying. Among those who flew to the site were Hare, Strand, Joe, and representatives from BC Parks. The weather only allowed for them to have two hours at the site.

As Strand walked around the glacier looking for items that belonged to this man, she wasn't sure that what they were doing was right. "I had a lot of turmoil about it." As at the Thandlät ice patch, she was hanging back from everyone else when she spotted a crumpled piece of material and picked it up. It was a woven spruce-root hat worn by the person who died there, although Strand didn't know it at the time. She bent down and picked up the misshapen and crumpled material, and she thought, "This is right. This is right." She felt a connection to this spot, the man who died here, and his belongings—as if he were reaching forward in time asking for his story to be told.

A day after Hare, Strand, and Joe confirmed that the site of the body was indeed archaeological and an extraordinary cultural find, CAFN leaders held a retreat in a remote roadless area on Kusawa Lake, just a half-day hike from the Thandlät ice patch. Their purpose was to discuss treaty issues with a neighboring First Nation. The retreat had been scheduled before the find, but the timing was fortuitous, and the body was an important topic of discussion.

"There were certain things we knew that we had to do. We knew we had to name him and that it had to be an Indigenous name. We knew that we had to follow what our Elders wanted," said Strand. CAFN leaders named him Kwädąy Dän Ts'ìnchį—which, in Dákwanjè, means "Long Ago Person Found." At the retreat, community leaders who included a Council of Elders decided that the body should not be left on the glacier, not necessarily because of its archaeological importance, but because this person was someone's relative—a father, brother, son, or uncle. In Athabascan and Tlingit cultures, if someone dies in your traditional territory, then it is your responsibility to care for that person, regardless of the deceased's cultural background.

While on retreat and unreachable, archaeologists in British Columbia were anxious to recover the remains; they worried that the partially exposed body might be scavenged by animals. When University of Alberta forensic archaeologist Owen Beattie examined photos of the body taken a few days earlier, he thought he saw evidence of scavenging. Beattie and British Columbia archaeologist Al Mackie drove to Whitehorse hoping that there would be a recovery effort in the next few days.

By this point, the media had caught wind of the discovery. Just as they had been with Ötzi, they were desperate for information. Dozens of journalists and television crews camped out near Kusawa Lake to catch Strand, Joe, and anyone who'd talk to them about the body as they came out of the retreat. "There were cameras and media everywhere," Strand told me.

When CAFN leaders and Elders returned from the retreat, they discovered that Al Mackie had been issued a ministerial order from the British Columbia government to recover the remains. CAFN leaders were concerned because this felt like an overreach. The battle over the Ancient One's remains was at its height. The relationship between the British Columbia government and CAFN regarding the Tatshenshini-Alsek region strained under the weight of decisions about the area that didn't include CAFN leaders.

The land that makes up Tatshenshini-Alsek Park had been set aside without any consultation with the Champagne and Aishihik First Nations. The Park's designation was a response to proposed mining developments in the heart of the region. While the intentions of protecting the park land were admirable, the failure to include the land's original inhabitants in the decision was infuriating. And then to add insult to injury, the British Columbia government included the word wilderness in the park's name. This was part of CAFN's homeland and had been for many thousands of years. The word *wilderness*, with its inherent connotation of the absence of people, disregarded their histories.

For years the people of CAFN had fought, without success, for a land claim agreement covering their traditional lands within British Columbia. Although their land claims in British Columbia are still outstanding, the CAFN government successfully negotiated co-management of the park. It was an important win.

One of the first things CAFN leaders did was drop the word wilderness from the park's name. They also argued for exclusive naming rights and, most importantly, sole responsibility for the interpretation of their culture and history. It was a well-timed shift in park management and meant that the CAFN government had control over Kwädąy Dän Ts'ìnchį and his belongings. He would not become another Kennewick Man.

The goals of traditionally trained western archaeologists and those of traditional Indigenous archaeologists and heritage managers don't always overlap, but where archaeologists from British Columbia and the First Nations found common ground was in figuring out who Kwädąy Dän Ts'ìnchį was. What culture did he belong to? How did he die and when did he die? To answer these questions, they'd need to study his remains and belongings. Under a special agreement, CAFN leaders agreed to a fifteen-month term of study. They issued a call for proposals to study Kwädąy Dän Ts'ìnchį and his belongings. Various teams of scientists who submitted proposals hoped to put together a detailed story about his life, glacier travel, trade, and relations between coastal Tlingit and Athabascan and interior Tlingit cultures before European colonization. In the meantime, Hare began sorting through the thousands of artifacts he had.

SEVEN

Long Ago Person Found

By the end of the 2003 field season, Hare had enough material to return to the question of when Indigenous hunters in the Southern Lakes area switched from the atlatl to the bow-and-arrow. "We had dozens of artifacts, so we started separating them into categories such as darts or arrows," said Hare. Diagnostic features of an atlatl included a dimple at one end of the shaft that would have fit into the spur on the throwing board of an atlatl. Another clue was the larger-diameter slotted hafting element where the point was attached to the shaft. And the third characteristic of an atlatl was a mid-shaft measuring longer than about forty inches.

Often objects found next to each other dated to the same period, and Hare literally fit the pieces back together. In several cases, the wooden shafts were still hafted to their stone points. Sometimes he'd find the missing pieces of an atlatl or an arrow many years apart. He had so much dateable material that he enlisted the help of three radiocarbon-dating labs. Some samples were sent to the IsoTrace Laboratory in Toronto. Others were sent to Beta Analytic in Miami, and still others were shipped to Lawrence Livermore National Laboratory in California. Once Hare laid out the chronology of the hunting tools based on the radiocarbon dates returned from the three labs, he noticed a striking pattern. "Everything younger than 1300 years old was entirely bow-and-arrow technology, and everything older than 1300 years was almost entirely throwing-dart technology," Hare told me.

For the first time, archaeologists were able to pinpoint with great accuracy the transition from one technology to another. This type of discovery is virtually unheard of in traditional archaeology, but with the sheer volume of organic materials and continuous advances in radiocarbon-dating techniques, these kinds of insights will probably become more common as archaeologists begin to sift through the artifacts they've collected from ice patches.

While it's hard to argue with the radiocarbon dates, transitions in tool use are tough to nail down because, said Hare, "you never assume you found the last example of something before it fades into history, and you never assume you found the first example of something." And yet the timing of the transition has held up over the last twenty-five years of surveying and dating all identifiable organic objects.

Hare says that the shift in technology occurred right around the time of a major volcanic eruption that originated west of the Southern Lakes region. Mount Churchill is an ice-capped stratovolcano in the Saint Elias Mountains in eastern Alaska. During the winter of 847 CE, Mount Churchill blanketed the southern tier of the Yukon in a layer of ash nearly a foot deep in places. It's known as the White River Ash layer. The eruption was so large that traces of the ash have been found in western and northern Europe.

Falling ash would have killed the plants that caribou and other large herbivores eat. Sulfuric acid (a byproduct of volcanic eruptions) is especially deadly to lichen. If large mammals were temporarily forced out of the area, people would have followed. If the eruption forced people out of the southern Yukon, perhaps they met people who were using bow-and-arrow technology and brought it home when caribou returned. It's a theory that is supported by a shift in caribou genetics that occurred right around 1200 years ago.

Tyler Kuhn, a biologist with the Yukon government, compared ancient caribou DNA collected from ice patches to modern caribou

DNA as part of his master's research at Simon Fraser University in Vancouver. "We used a variety of materials from the ice patches," said Kuhn. "Fecal matter, bones, teeth, antlers, essentially whatever was collected that had a date associated with it. That's the key part. We needed to know how old things were, so if we had a sample that we didn't know how old it was, it wasn't going to help us understand the story."

The DNA collected from ice patches told a story of genetic continuity, or one population interbreeding and living in the Southern Lakes area, until right around the time of the eruption. "Between five thousand years ago to about the time of the eruption, the genetics are very similar between what's found in the Southern Lakes area and everywhere else. And then after that time, there's just a subtle difference between what's found in the Southern Lakes and everywhere else," said Kuhn.

Kuhn told me that the shift in genetics could have been caused by one of two things. Initially, he focused on the idea of the displacement of a large mixed genetic group of caribou and recolonization by a smaller group that had less genetic diversity. "So, in genetics speak," said Kuhn, "we're talking about founder effects." The founder effect occurs when a small group of individuals with essentially randomly selected genetic information from the larger parent group arrive in an area and they start to breed. Over time, their genetics evolve slightly from the main population. "But," he said, "you could accomplish the same thing by having a large group whereby only a few individuals survive a dramatic event like a volcanic eruption. The population goes through what's called a bottleneck, and those animals have less genetic diversity than the larger group, which then changes a bit over time." Outside of this eruption bubble, caribou don't show the same interruption in their genetic code.

And although the eruption of Mount Churchill, genetic changes in Southern Lakes caribou, and the abrupt and complete abandonment of the atlatl in favor of the bow-and-arrow all occurring at the same time paints a compelling picture, there could be other reasons for the shift in technology. For one, right around the time of the eruption, people had just modified the design of the bow-and-arrow and as a result, it became much more efficient. Maybe it just wasn't worth the effort to Southern Yukon peoples to learn how to use a new hunting tool until then.

More Than Caribou Hunting

There are now forty-three artifact-bearing ice patches that are part of a regular monitoring program. The transition from a rescue mission to monitoring has freed up some time for more in-depth analysis of the hundreds of hunting tools already collected. "We've now started comparing arrows found at ice patches to the ethnographic descriptions of arrows, because it's a lot more telling of what's going on in people's minds when they are up there hunting in the mountains," Yukon government archaeologist Christian Thomas told me.

Thomas has been involved with the Yukon ice-patch project from the beginning. He helped survey ice patches and extract ice cores as an intern during the early years of the project. He returned in 2007, a few years after earning his master's degree in archaeology from the University of Alberta. When Hare retired in 2017, Thomas assumed his role as the Senior Projects Archaeologist for the Yukon government.

Thomas says that the Yukon ice-patch project is led by the First Nations. They decide not only what gets done but also how it's done. It's part of a long path toward reconciliation between First Nations people and western archaeologists. The Yukon government has the resources to keep a project like this going, but it's the different First Nations that decide on what they'd like to know more about. These artifacts, after all, are part of their history—a history that was partially and purposefully suppressed by colonialism. One of those Indigenous-led projects is to look more closely at the hunting tools retrieved from the ice patches.

Thomas began sorting the arrows by differences in the species of wood used and point type—bone or stone. At the Yukon ice patches, Hare and Thomas only found arrowheads made of bone, but they knew that stone arrowhead points were relatively common in the Yukon and elsewhere. By teaming up with archaeologists in the Northwest Territories, they were able to map out a more complete story of arrow design. They found that there were two main types of arrows. Some were made of small stone points hafted to birch shafts, which measured more than two feet long. The second type were made of antler points, some of which were barbed along one side and hafted to a spruce shaft measuring no more than about two feet long.

But when Thomas laid out the arrows chronologically regardless of type, he didn't see an obvious pattern. In other words, the materials and construction techniques—like the length of the shaft, species of wood used, and size and type of the arrowhead—varied over time but in no discernable way. It wasn't as if heavy stone arrowheads were used up to a certain point and then hunters switched to a smaller arrowhead afterward.

Thomas wondered why two point types were used at the ice patches at the same time. To make sense of it, he began reading accounts of hunters collected by ethnographers during the last century and talking to Indigenous hunters while out surveying ice patches. "What stuck out was that the First Nations had these arrows for every occasion. They had a different type of arrow for hunting birds, and they had a different arrow for hunting large game," said Thomas. And then it dawned on him that perhaps the differences in arrows found at the ice patches had to do with targeting different animals.

"We'd only been talking about these sites as caribou hunting sites. But there are a lot of sheep up there," said Thomas. And in fact, the Thandlät ice patch and Kwädąy Dän Ts'ìnchį had both been found by sheep hunters. When Thomas tallied the bones from the ice patches, he discovered that two thirds of the collection were made up of caribou, and the rest were from sheep. So Thomas thought: What if one arrow type was for hunting sheep and another was for hunting caribou?

It was an *aha* moment for Thomas.

Although he'd found references to sheep-hunting in the ethnographic record, any discussion of the types of points used for sheep versus caribou was left out. But he did find one source that said when greater penetration was necessary, hunters would use a stone point. Sheep spook easily. But caribou are very curious, and so hunters are able to get much closer to them than they can to sheep. "The caribou just come right up to you," said Thomas. Their curiosity is probably what made them so amenable to domestication in other parts of the world.

A stone-tipped arrow with a birch shaft flew farther and faster, which meant the hunter didn't need to get as close to a sheep to make a successful shot. But these arrows may have been less precise in hitting their targets, which is why hunters also used spruce arrows with barbed antlers when hunting caribou. These were more precise but required closer proximity to the target. In other words, hunters selected the best tool based on what was needed in a particular hunting situation. Their shape and the size of the arrowheads and different parts of the arrow resulted in tradeoffs between precision, shooting range, and depth of penetration when an animal was hit. "Differences in hunting sheep versus caribou represent a different hunting problem that got turned into a design problem," said Thomas. That problem's solutions may also have involved changes in strategy, not just arrow design.

Thomas says that hunting-tool traditions changed based on people's needs, but it's hard to get into the mind of someone who lived a thousand years ago. "I think that a lot of the new insights will come from just talking to people about hunting in the mountains. It's not only about documenting and preserving these objects, but making meaning from them. These artifacts are the ancestors of present-day hunting ideas," said Thomas.

Stone Hunting Blinds

Some of these new insights may be found by looking more closely at hunting blinds. Kelsey Pennanen, a PhD student at the University of Calgary, is taking a closer look at stone features found near ice patches. The project was initiated by the Kwanlin Dün First Nation, who wanted to understand the broader subsistence landscape.

Carcross/Tagish First Nations Elder Art Johns learned of two sets of stone blinds as a child. "Art had been a highly skilled and professional big-game outfitter," Hare told me. "He had been up in the mountains riding horses and guiding hunters from the time he was about sixteen." The first is a set of seven blinds situated atop a ridge between two ice patches called Friday Creek and Alligator. The ice patches are separate now but were probably once connected. Hunters would have waited behind the stone blinds and harvested animals either exiting or entering the ice patch from the ridge above. The second set of blinds lies across the valley from the Friday Creek ice patch. The blinds face south toward the ice patch. Their arrangement suggests that animals on the ice patch may have been driven to hunters waiting behind the blinds across the valley.

These blinds are hard to see. Some of them have toppled over, but more than that, the configuration of blinds on the landscape with respect to ice patches, game trails, and other landscape features (such as ridges) is hard to take in from a horizontal perspective. They are best viewed from above.

Pennanen uses digital technologies to record archaeological sites. "I've been using drones and UAV [unmanned aerial vehicles] photogrammetry to reconstruct the landscape in three dimensions—size, shape, and orientation," she told me. Pennanen grew up in northwestern Ontario. "I've always loved archaeology as a way to read the layers of earth, like the pages of a book through time." She wondered how she could use her skills to provide the information that Kwanlin Dün and

145

other First Nations in the southern Yukon were interested in, especially for sites that are considered at-risk.

Obviously, ice patches are at risk of melting. But surface features like cairns and blinds are also at risk of crumbling naturally or of being destroyed. People don't always recognize that someone must have made them, or if they do, they think they were built recently. With greater recreation in the mountains, some of these features have been used as mountain-bike jumps or are toppled over by careless recreationalists. While education goes a long way to discourage people from disturbing these features, capturing the features digitally is its own sort of preservation. "First Nations can then use these digital models to create their own interpretive displays," said Pennanen.

While some of these features were mapped in the process of conducting ice-patch surveys, many of them were found by First Nations hunters in recent years. Keith Wolfe Smarch, Tlingit master carver, found a large complex of stone blinds while sheep hunting. "One night I took a shortcut back to the road and my truck. I was packing sheep meat and I was coming down the mountain. It was just getting dark, and I thought it was a hole in the rock, that maybe a glacier had created it." He didn't think much of it until a year or two later when he found another one in the same spot, but it wasn't until his third visit that he realized that these weren't boulders pushed around by a glacier; they were made by people.

Smarch came back to the site later and found fifteen additional stone hunting blinds in the same area. "Each time I walk back there, there's a band of sheep there." But he thinks that the blinds were actually made for hunting caribou rather than sheep. "And there are trails that wind all through these hunting blinds. When I realized what they were, I thought to myself, 'I'm probably an ancestor of the person who used this blind.'"

He took Pennanen to the blinds and she mapped them from above with a drone. Using specialized software, she stitched the hundreds of images together and then played around with shading and coloring to see how well the blinds stood out in the imagery. When she looked at the digital images, the semi-circular shape of the stone blinds was strikingly obvious. Many of them are situated on either side of a game trail that was invisible while standing at the site.

I asked Pennanen if she knew how old the blinds were, and she gave me an unexpected answer for an archaeologist. "To me, that's not where the value lies," she said. "It's about the fact that we have a nine-thousand-year hunting tradition that continues to this day. These blinds are still being used. So, if a hunting blind or cluster of blinds had a certain configuration three thousand years ago, who's to say that it wasn't a modification from six thousand years ago?" Or, for that matter, a modification in the present day by a hunter like Smarch?

Smarch has used these hunting blinds himself. He said sheep have good hearing and eyesight and a keen sense of smell, so it can be hard to stay hidden. "Once you learn how to hunt, for me it's pretty simple. I get really close to them. There's a few tricks that I know. You gotta stay down and be quiet. You gotta be careful of noise because noise carries a long way in the mountains." If the wind isn't blowing in your favor, for example, his uncle taught him to light some grass on fire to help disguise his smell (sheep are used to the smell of smoke from forest fires). A hunting blind provides a good fire ring. But these stone structures may have been used for purposes other than hunting.

When Pennanen began visiting these blinds with Indigenous hunters, she realized that she'd been thinking of them in only one way—as concealment during a hunt. But there are others located too far away from ice patches or any game trail to have been used for hunting caribou or sheep. At one site, local Indigenous hunters told her that the blinds overlooking a broad valley were probably used by hunters to sit and watch the herd, how they moved, where they were going, and to gain insight into herd health. In other words, they were scouting blinds. Other hunters told her that some were used as windbreaks just as often as for concealment. Maybe some of them were used as shelters when the weather turned and the hunter couldn't make it back to camp. Lawrence Joe (former Director of Heritage, Lands and Resources for CAFN) told me that some of the blinds were probably used as meat caches. These were insights that Pennanen admits she wouldn't have had on her own.

"I shouldn't be the one interpreting these features," she said. That's up to the Kwanlin Dün, Carcross/Tagish, and other First Nations she's creating these models for. She considers her role as an archaeologist to be a conduit: She uses her set of skills to provide information to the First Nations based on their needs. "For so long we've had the voice and been the interpreters of the past. To take over an entire continent and really colonize it, you need to intentionally disregard the histories of those who were here before. And archaeology plays a pretty big part in that," said Pennanen.

To protect these features and get the word out, the Kwanlin Dün plan to make some of them into interpretive spaces. The Champagne and Aishihik First Nations are interested in education and outreach. But all of them talked about the cultural connection to these landscapes and bringing that back, especially for Indigenous youth. Pennanen hopes to create a three-dimensional experience where people can immerse themselves in the landscape as it exists now, because these ice patches may not exist in the future. "Archaeology is only relevant to the present. It's really about how we can help people today create and maintain these relationships to the caribou and to the landscape," said Pennanen.

All the artifact-bearing ice patches are within a day's walk of a historically known village or seasonal camp, often located along the shores of lakes full of fish. While caribou and sheep were important subsistence animals, there are many others that were hunted that don't have the same relationship with ice. So, ice patches don't tell the whole story of hunting traditions—just a part of the story of not only hunting but of a cultural landscape in which people harvested berries, fished lakes and rivers, and made clothing and baskets from the natural materials found around them. For example, ground squirrels were often trapped near or at ice patches using snares. Their skins were stitched together to make clothing and their fat was rendered and mixed with berries and other foods to eat during lean winter months.

Western archaeologists focus much of their attention on objects and what those objects reveal about culture, Indigenous archaeologists and cultural heritage managers have a different relationship with artifacts. Choctaw archaeologist Joe Watkins writes that "[western] archaeologists tend to focus on the physical, technological, or esoteric attributes of an artifact, while Indigenous populations tend to focus on the ritual or social importance of the artifacts." It's so obvious that it's easy to forget: There was a person on the other end of these artifacts. They were made by people—people just like us. And nothing brings that home more than finding the person to whom these things belonged.

Indigenous Stories of Glacier Travel

Scientists had less than two years to study Kwäday Dän Ts'ìnchị and his remains. Although the sanctioned research period was brief, from those fifteen months of research emerged a story nearly as detailed as Ötzi's. Kwäday Dän Ts'ìnchị was found along the northern margin of a small unnamed glacier in what is now considered a remote location, but not so long ago it had been a trading route between the traditional villages of Klukwan and Klukshu. Kwäday Dän Ts'ìnchị died during the final stages of the Little Ice Age, sometime between 1720 and 1850. Because he had been trapped in a glacier and not in a stationary patch of ice, he wasn't nearly as well-preserved as Ötzi even though he was much younger. Fortunately, his body had been partially protected by three large, knuckle-like nunataks rising out of the glacier. The nunataks, which are isolated mountaintops that protrude above an ice field, both slowed and redirected the flow of ice just enough so that most of Kwäday Dän Ts'ìnchị's bones, skin, hair, and organs—along with some of his clothes and tools—were intact.

How he died remains a mystery, but it's unlikely he fell into a crevasse. He was too well preserved to have endured such a fate. Although he was a fit and healthy man around the age of twenty, mountain travel is inherently risky. Aside from the dangers specific to glaciers, weather in the mountains is capricious. Snow can fall in any month of the year. Hypothermia even at the height of summer is always possible.

Trade routes between the coast and interior often included crossing glaciers. These are some of the most glaciated mountains in the world today and would have been even icier during the Little Ice Age. During the Little Ice Age, glaciers flowed across valleys, sometimes merging with glaciers flowing down from the mountains on the opposite side. These glaciers blocked rivers and created large lakes behind them that would sometimes burst, destroying everything in their path. Some traditional stories recount people traveling under glaciers rather than over them.

Deikinaak'w, a Tlingit man, said in 1904 that "in one place the Alsek River runs under a glacier." Rivers would wear away the ice and create bridges like the one in this story. "People can pass beneath in their canoes, but if anyone speaks while they are under it, the glacier comes down on them. They say that in those times this glacier was like an animal and could hear what was said to it."

English travel writer and journalist Edward Glave told a similar story about the Alsek Glacier during his late-nineteenth-century visit. It "reaches from the slopes of these mountains and trends away to the river, where its intruding walls of ice are torn asunder by the angry torrent and carried away to the Pacific Ocean."

Those who traveled across glaciers did so out of necessity. While dangerous, glaciers were often easier avenues of travel than through thick Alaska bush at lower elevations. Strand told me that there are many songs about crossing glaciers—when you cross and when you don't cross. And there are many songs about glacier death. Because of these stories, local Indigenous Elders weren't particularly surprised by the discovery of Kwädąy Dän Ts'ìnchį.

After the discovery, Strand and her counterparts with other First Nations began gathering stories of glacier travel from within their own communities. Many stories were already known and documented, so they started with their archives. Elders also came forward with stories they knew of people who had gone missing while traveling from the interior to the coast to trade with the Tlingit. Kwädąy Dän Ts'ìnchį's relatively recent death gave Strand hope that one of these stories might be about him.

Strand's grandmother, Mrs. Annie Ned, chronicled a story of a coastal Tlingit man who fell through a crevasse on a glacier near Kusawa Lake while on a trade mission. He had been traveling to the coast from the interior with his trading partner when he fell into a crevasse around the time when Kwädąy Dän Ts'ìnchį died. His trading partner thought the Tlingit man was surely dead, so he continued to the coast to tell his partner's clan of the tragedy. They held a funeral potlatch and then several men traveled back to the glacier to retrieve the body, but they discovered that he had somehow survived by wrapping himself in beaver skin and moose hide. They managed to rescue him by lowering someone down with a rope and pulling him back out. He'd been lost in the crevasse for ten days. Although half-starved and hypothermic, that man lived, so the story wasn't about Kwädąy Dän Ts'ìnchį. In the stories I read about Indigenous glacier travel in the Yukon, survivors always attempted to recover the bodies, but that wasn't always possible. Crevasses can be hundreds of feet deep.

One such story provided a glimmer of hope that Kwädąy Dän Ts'ìnchį's identity might be revealed. Champagne and Aishihik First Nations Elder John Adamson recalled hearing about a man who had gone missing in precisely the same area where Kwädąy Dän Ts'ìnchį died. When he brought it up years after the body was discovered, it had been sixty or seventy years since he first heard the story, and his memory was hazy. He brought up the story again just months before he died, but no more information was discovered. With the passing of time, the probability of discovering Kwädạy Dän Ts'ìnchị's name grows dim.

To the Indigenous cultures who have lived in frozen landscapes for thousands of years, glaciers are much more than bodies of ice. A glaciated landscape is a landscape that listens, has a sense of smell and hearing, is quick to anger, and responds to indiscretions with vengeance. Stories about glaciers convey the consequences of improper behavior around them. One convention prohibits cooking with grease near a glacier.

The Saint Elias Mountains are home to one of the highest concentrations of surging glaciers in the world. Surging glaciers are moody. They flow unhurried for decades, and then, without warning, they rage forth at many times their normal speed. Scientists don't completely understand how or why glaciers surge, but it seems to have to do with meltwater forming between their base and the ground, which then causes a glacier to slip along its bed. Climate change may be making surges more common. As the climate warms, meltwater pools on a glacier's surface. The water eventually finds its way through the glacier and accumulates at its base. This extra meltwater unmoors the glacier, resulting in a surge.

Although some surges happen instantaneously, most surges are too slow to observe with the naked eye. In early winter 2020, Denali National Park's thirty-nine-mile-long Muldrow Glacier shifted from flowing less than a foot per day to flowing as much as sixty feet per day. The glacier became riddled with cracks. Normally, the Muldrow Glacier is a safe access point to climb Denali, but the surge forced the closure of the route. It finally stopped surging in midsummer 2021. While this surge lasted less than a year, they can persist up to a decade.

Perhaps there is a connection between the forbidden act of cooking with grease while traveling across glaciers and the abundance of surging glaciers in Alaska and the Yukon. Outside of the Saint Elias and Alaska ranges, glaciers in western Greenland, Svalbard, Iceland, and the Karakoram, Pamir, and Tianshan mountains contain surging glaciers. But while these types of glaciers are found in dozens of mountain ranges worldwide, only 1 percent of the world's 215,000 glaciers surge.

Kwädạy Dän Ts'ìnchị's Final Days

Although Kwädạy Dän Ts'ìnchị's real name may never be known, his clothes, tools, and body tell the story of his last few days and shed some light on his earlier life. His journey began near the coast sometime in late summer. He had eaten a meal of salmon, sea urchin, and beach asparagus (his clothes had been covered in its pollen) before setting out on his journey. Beach asparagus (aka sea beans, pickleweed, glasswort, and samphire) is an aquatic succulent that grows within a narrow range along intertidal areas of coastal salt marshes. The green stems tipped in red are harvested throughout the summer, but before they finish flowering and take on a woody texture. Based on the distribution of beach asparagus, he likely started his journey at Klukwan, near the Chilkat River estuary. He was raised along the coast but moved inland during the last year of his life; perhaps he married someone from the interior and lived with her family.

Oral histories tell of three major trails from the coast to the interior. One begins near the Chilkat River estuary in the village of Klukwan, at the head of the Lynn Canal. From Klukwan the trail branches northeast along the Klehini River to Shäwshe. A second trail beginning at Klukwan heads north to Kusawa Lake. The third trail connecting the coast to the inland lies farther north at Dry Bay and heads northeast along the Alsek and Tatshenshini rivers to Shäwshe. Kwädạy Dän Ts'ìnchị was found between the two trails leading to Shäwshe.

The vastly different environments of the coast and the interior meant that each group of Indigenous people had unique and important items that the other group wanted. The coastal Tlingit traded dentalia shells, obsidian, dried seaweed, fish oil, spruce-root baskets, cedar boxes, fungus for paint, medicinal roots, and native tobacco. In return, the Southern Tutchone traded raw copper, sinew, mountain-goat wool, tanned moose and caribou hides, and clothing.

Kwädąy Dän Ts'ìnchį traveled an average of twenty miles per day over two or three days, following established trails through thick coastal rainforest. The Klehini River braids its way northwest through a densely forested valley of Sitka spruce and western hemlock with thick shrubby undergrowth. Gravel bars created by flooding and shifts in the river's course would have made travel easier, as it scoured the river of plants.

As he gained altitude toward the headwaters of the Klehini River, mountain hemlock and alder replaced Sitka spruce and western hemlock. He then headed north where he could have either stayed along the river channel or climbed into the alpine atop Copper Butte and then down a valley toward a place called Mineral Lakes, where he stopped for a drink and a snack of moose or bison meat. He was just three miles from where he died. Not long after his break, he gained his first views of the Samuel Glacier. From Mineral Lakes, the most direct route to the glacier where he met his end is across a three-mile-long southern lobe of the Samuel Glacier.

Kwädąy Dän Ts'ìnchį traveled light. Along with the knife found by the three hunters, he carried a robe made of Arctic ground-squirrel furs stitched together in a beautiful patchwork of ninety-five pelts. It had been repaired with humpback and blue whale sinew. This was the pile of sopping wet material that Hanlon and his companions saw. The robe he carried with him is of Tutchone origin and was likely a prized item. The hat that Strand found was made of woven spruce roots, probably Sitka spruce, which was widely used in weaving by the coastal Tlingit. It was shaped like a truncated cone with an inner headband that would fit comfortably on his head and secured with a hide chin strap. He also carried a beaver-skin bag made from a single piece of fur folded in half with the fur on the inside. The bag was stitched up the side with moose sinew and may have been folded or tied at the bottom to hold its contents. One of the most important items was a leather pouch that CAFN Elders identified as a medicine pouch. No one knows what he carried inside it, since custom dictates that the contents of medicine bags are personal and private.

It was important to CAFN leaders and citizens that Kwädąy Dän Ts'ìnchį and his belongings remain under CAFN's control and not be accessioned into the museum's collection, where they'd wind up on display or in a dark storage room. Instead, Kwädąy Dän Ts'ìnchį and his belongings were loaned to the museum for temporary storage and study. But CAFN never claimed ownership of Kwädąy Dän Ts'ìnchį. Rather, they claimed responsibility for him and his belongings. They wanted to be sure that this person, whoever he was, was treated according to Indigenous customs.

And so, on July 18, 2001, nine people flew by helicopter to the discovery site. They carried with them a hand-carved wooden box containing Kwädąy Dän Ts'ìnchį's ashes. While his hat, beaver-skin bag, robe, and other objects found at the site were retained, the medicine pouch was cremated along with his remains.

It was a good day for flying. Patchy clouds blew over high in the sky. The area was still covered in last winter's snow. They built a cairn under which they laid the ashes. The ceremony was short and simple, with a song and a few words of respect, and then Kwädąy Dän Ts'ìnchį was laid to rest, right where he was found two years earlier.

No Such Thing as Wilderness

In 1994 Tatshenshini-Alsek Park, where Kwädąy Dän Ts'ìnchį died—along with Kluane National Park and Reserve (Yukon), Glacier Bay

National Park and Preserve (Alaska), and Wrangell–St. Elias National Park and Preserve (Alaska)—was designated a UNESCO World Heritage Site. Together, these parks represent the largest protected area in the world. But when Kwädạy Dän Ts'ìnchị died, there was no such thing as a "wilderness" apart from where people lived, at least not in North America. Tatshenshini-Alsek Park only looks untouched; it hasn't been that way since people first arrived on this continent many thousands of years ago.

The word *wilderness* was removed from Tatshenshini-Alsek Park's name at the insistence of those whose relations there go back many thousands of years. But the criteria under which UNESCO designated this enormous region as a World Heritage Site were natural ones, including for the "outstanding beauty of glacially carved landscapes, salmon-spawning rivers, grizzly bear habitat, and home to the rare blue phase of the black bear known as the glacier bear." While all these things are true, it is also an ecosystem of human relationships with the environment.

In *Wilderness and the American Mind*, Roderick Frazier Nash writes that Americans built civilization from the raw materials of the wilderness, and from the built environment, people created the idea of wilderness. Before so-called civilization, there was no need to distinguish between the built world and wilderness, because the land was all wilderness. French naval officer Jean-François de Galaup, comte de Lapérouse, wrote in the eighteenth century that the wild and pristine places he encountered were "on the verge of the world." When William Hanlon came upon the broken stick at the edge of the glacier he remarked, "well, I guess we're not the first to hunt here." In a landscape as wild and rough as this park, it only feels as though no one has been there before. Mountains are the last bastion of "wilderness" as we think of it today, but it was Europeans who brought with them the idea that nature is not only inanimate but also separate from culture. And it's part of why the human connection to mountain landscapes was overlooked for so long.

156

A Story Beyond Hunting

Since that first ice-patch artifact was discovered in 1997 on the Thandlät ice patch, Thomas says that in the last few years caribou have returned. "It was 2018 or 2019 and we were coming in for a landing and I said, 'Oh, my God, there's a caribou,' and it was running. And then I saw a grizzly bear jump out and start chasing it. I don't know what happened because we decided not to land. And I just remember being there in 1998 and there was all this caribou dung but no caribou."

All three Southern Lakes herds have grown, and there are now an estimated four thousand caribou among the three herds, up from the 1500 animals at the start of the ice-patch surveys. Despite the increase in the Southern Lakes herds, Kuhn says that hunting of caribou is still prohibited. Although direct human pressures like habitat fragmentation, roads, and mining continue to threaten caribou, their biggest threat is climate change. Warmer summers and winters mean that the ice—glaciers, ice patches, and ice caps—is melting fast. For cold-adapted animals like caribou, a warming climate is their primary threat.

"Probably the worst years of melt since the first couple of years [of the ice-patch project] was in 2018 and 2019. We had a combination of low winter precipitation and very, very hot summers," said Thomas. In 2019 all the seasonal snow atop the ice patches had melted. He visited a few ice patches that he hadn't been to in five years, and they were gone. Thomas thought for sure that things were going to start disappearing after that. But since then, they've had three winters with incredible precipitation. "We've never seen snow in our mountains around White-horse and Haines Junction like we have in the last three years" (the winters of 2019–20, 2020–21, and 2021–22). "We're actually seeing cor-nices in the dead of summer at these ice-patch areas, and nothing's melting out except for at two sites," he told me.

But a couple of good years don't amount much when you look at the climate over the long term. Average annual temperature in the

Yukon over the last half century has increased by 3.6°F (2.0°C). This temperature increase is more than double the global rate, especially during winter, when the average temperature has increased by 5.8°F (3.2°C). These temperatures are the warmest of the last twelve thousand years. There aren't weeks-long periods of below-zero conditions (to -40°F [-40°C]) like there used to be. This isn't a scenario of projected future climate change: It's happening now.

And this is the paradox. Ice patches only reveal artifacts during high-melt years—when less snow accumulates than melts. It's why archaeologists survey in late summer after most, if not all, of the last winter's seasonal snow has melted but before the first snow of autumn covers the artifacts again. That window of opportunity is narrow but growing.

Thomas says, "when we think about climate-change archaeology, things are being lost, and we must do something. But the narrative never goes back to what it is we're losing." This is a record of an exceptional hunting tradition, and that is what's at risk. People want to maintain that tradition and keep connected to it despite all these changes. And that is what these artifacts preserve. "It's not just straight ballistic engineering. Even the gun owners out there have engravings on their gun and special features. It's not just a gun. And there's a lot of art on these artifacts," says Thomas. Eagle feathers, the bright red shaft of a northern flicker feather, and the plumage of white-tailed ptarmigan, gyrfalcon, and short-eared owls were used for feather fletching on arrows. "So, you're seeing not only something that was used for hunting, but you're seeing a story there beyond the hunting on the mountain."

As the oldest ice begins to melt, the artifacts being found are getting older and older. It's exciting on the one hand to think about what might melt out next (and where), but it is a bittersweet kind of excitement. Every artifact adds to the story of our deep-time relationship with ice while at the same time reminding us of what we are losing, as Thomas pointed out.

A decade after the Yukon ice-patch project began, Montana archaeologist Craig Lee made a stunning discovery while surveying an ice patch in the mountains outside Yellowstone National Park, near where I live. Eight inches of bighorn sheep and bison dung blanketed the ice patch, giving off an overwhelming aroma of decay. The artifact that Lee found would become the oldest to have ever melted out of an ice patch—a title the object still retains.

EIGHT

The World's Oldest Ice-Patch Artifact

In late summer 2007, archaeologist Craig Lee was surveying a patch of ice near Yellowstone National Park when he noticed what looked like a small branch poking through the snow. He hesitated; he wasn't supposed to be there. Although he stood on public land, he had crossed the state line from Montana into Wyoming. The permit that gave him permission to survey and collect artifacts only applied to Montana. He knew he stood in Wyoming, but this one ice patch looked so promising. He had seen it from a distance and had always been curious about it. Besides, he reasoned, it was such a classic-looking ice patch that he thought he'd at least get a photo of it for future grant proposals. Before he knew it, he'd arrived at the edge of the ice patch and was staring at what was potentially the discovery of a lifetime.

Lee is an archaeologist at Montana State University in Bozeman, where he lives with his wife and teenage daughter. In 2001, Lee started doing ice-patch research on the Alaskan side of the Wrangell–Saint Elias Mountains as a graduate student assistant at the Institute of Arctic and Alpine Research (INSTAAR) in Boulder, Colorado, under the tutelage of E. James Dixon. This was just a few years after Greg Hare took up ice-patch archaeology work on the east side of the range.

While glaciers in these mountains are well mapped, perennial ice patches are often overlooked in inventories of the cryosphere. Ice patches are difficult to map because they are small and occur in complex terrain. To narrow down their search, Dixon and Lee teamed up with William Manley—a geospatial analyst also with INSTAAR. Manley used satellite imagery along with information about the range of Dall sheep, mountain goats, and caribou—animals that might have attracted hunters in the past. He also looked at ethnographic data on cultural trails and traditional hunting areas. Because of the remote location and short season for ice-patch archaeology surveys, the team used a helicopter to access the ice patches with the highest potential for artifacts.

Although Dixon and Lee identified more than 150 promising ice patches, only five yielded archaeological material during three years of surveys. But these were years of above-average snowfall, which meant that even if they visited ice patches with artifacts, they might not find any because they were still buried under last winter's snow. Or perhaps any artifacts had already decomposed when the ice patch was smaller in the past.

Still, they found the usual darts and arrowheads, some of which were made of wood, sinew, and antler projectile points. They also looked at a few glaciers near the ice patches. There they found Dall sheep and caribou bones, as well as bones from coyotes, birds, small rodents, and, strangely, a fish. Just how the fish got into the ice patch is a mystery. Something had obviously brought it there, but what they couldn't say.

Although the fish bones were an oddity, it's not all that unusual to find animal remains beyond the typical large mammals melting out of ice patches. Jørgen Rosvold, a biologist and assistant research director at the Norwegian Institute for Nature Research, has made a career out of studying animal remains from ice patches. He found them to be microcosms of life that go beyond a human-animal hunting tradition. The cold ice patches become death traps for insects and other arthropods. But ice is also habitat for ice worms, one species of which is entirely dependent on glacial ice. Ice worms feed off snow algae, which paints the ice in a soft pink. Insects and ice worms inevitably attract polar and alpine birds like horned larks, rosy-finches, snow buntings, white-winged snowfinches, glacier finches, and alpine accentors.

Predators also make use of ice patches. Wolverines often cache their kills in cold places, which keeps the food from spoiling until the wolverine returns days, weeks, or possibly months later. Some animals, like the brown bear and snow leopard, use ice as travel corridors much like people do. In 1926, a mountaineer found a mummified leopard on Mount Kilimanjaro. The discovery is said to have partly inspired Ernest Hemingway's short story "The Snows of Kilimanjaro." Over the last two decades, extinct mammals from the last ice age have emerged from the Siberian permafrost, including a wooly rhinoceros calf (ten thousand years old), a cave lion cub (thirty thousand years old), and a pair of wooly mammoth calves (forty thousand years old).

On the microscopic end of the size spectrum, viruses and bacteria have emerged from ice and have even been revived after laying dormant for thousands of years. In 2016, a heat wave in Siberia reawakened anthrax spores melting out of a reindeer carcass that had been preserved in permafrost for the previous seventy-five years. The bacteria killed a young boy and hospitalized dozens of others. Of the more than 2600 reindeer infected, 89 percent died. And in 2021, researchers from Ohio State University found genetic material from thirty-three viruses melting out of glacial ice collected from the Tibetan Plateau. Many of the viruses were previously unidentified. Some scientists think that melting ice could be the source of the next pandemic, especially as the oldest ice begins to melt.

Ice-Patch Archaeology in the Greater Yellowstone Area

The archaeologist in Montana with the National Forest Service heard about the Alaska project and asked Lee to survey some Montana ice patches. So, Lee wrote a proposal with the Forest Service and got approval and some money. To narrow down the possibilities, Lee looked at topographical maps combined with satellite imagery to try and find promising ice patches. Lee selected satellite imagery obtained during high-melt years—that is, years with above average temperature. If an ice patch was still present during a high-melt-year image, then Lee thought it might be perennial. He also looked for ice patches with a dark organic layer. He knew from his work in Alaska and from talking with Greg Hare that this was an important feature of artifact-bearing ice patches.

"The summer of 2006 had been pretty hot and dry and there wasn't much snow," said Lee. He had hiked to a place in the northern Greater Yellowstone ecosystem. At the edge of the ice patch there, Lee found two rooted Engelmann spruce stumps. Since the ice patch was above modern-day treeline, there must have been a forest there in the past. Radiocarbon dating of the tree stumps revealed that the trees died roughly eight thousand years ago, which was about five hundred years after the end of the Holocene Thermal Maximum. Temperatures were as much as 11°F (6°C) warmer than historic summer temperatures. When temperatures cooled and snow filled in the site, the trees died. But Lee didn't find any organic artifacts.

Then he headed to explore the high plateaus above ten thousand feet, dotted with dozens of kettle lakes scooped out by glaciers. At one perennial ice patch, Lee found a cut piece of wood at 11,250 feet. It was bent and not very well preserved. "It's something only an archaeologist would love," he said. Lee was disappointed. He'd been looking for things that were unequivocally made by people. It's not always easy to tell if an object was human-made, especially if it's broken. The cut stave dated to about 5350 years ago. The object hinted at past use of ice patches, perhaps as hunting sites as in Norway and the Yukon, though he couldn't be sure.

Unlike in Norway, the Yukon, and the Alps, there is no dedicated ice-patch archaeology program in the United States. And not because Lee isn't interested in developing one; there hasn't been the same level of support enjoyed by southern Yukon and Norwegian archaeologists. Many archaeological projects in the United States occur in relation to construction projects. Yet all permanent ice in the United States and probably almost all of the ice in Canada exists on public land. There are thousands of ice patches in the twenty-four-thousand-square-mile area in and around Yellowstone National Park. Lee says that he might be able to visit a dozen ice patches in a year if he's lucky. Vast remote roadless areas take time to survey and helicopters are expensive. So, Lee walks to most of his sites. "One day I might write a book about the peaks I've been near but haven't climbed," he joked. Ice patch sites are always below the surrounding peaks. By the time he climbs up to a potential ice patch, he's not that interested in "bagging the peak." He's got work to do. Besides, not every ice patch contains artifacts.

In 2007, Lee backpacked fifty miles out on a large plateau in the Absaroka Range. "Oh, it was horrible," he said. "There was no ice!" Even though he had looked at imagery and identified potential perennial ice patches before the trip, it turned out that none of them were perennial. It was one of those high-melt years, which normally are great for finding ice-patch artifacts—but only if you manage to find perennial ice patches, which brings us back to the Wyoming ice patch he hadn't intended to survey.

The Wyoming ice patch had a black rind of dung from bighorn sheep or possibly bison. When Lee got to the edge of the ice patch, he noticed bones everywhere. "It wasn't clear to me if those bones were anthropological," he said. Meaning, Lee wasn't sure if the bones were there because the animals they'd belonged to had been hunted at the ice patch or if they had been killed by a predator or died there for another reason.

As he wandered along the edge of the ice patch, he noticed a long, cut piece of wood, bent in the middle but otherwise intact. Although he didn't know how old it was just by looking at it, he suspected the object was of great age. But he was faced with a dilemma. "It's absolutely verboten to do archaeological research in a place for which you don't have a permit," he said. But as he looked around, he noticed a Doritos bag, boot prints, a broken pair of sunglasses, and parts of a snow machine. The site was already disturbed, and he wasn't all that far from a road. In about forty-five minutes, most people could hike to the site without too much trouble. There was so much trash he was afraid this irreplaceable object would be taken by someone who'd keep it in a personal collection or that it would get destroyed by weather or trampled by people or animals. So, he collected it.

As soon as he had cell-phone reception, he called the federal archaeologist with jurisdiction for that part of Wyoming. He was talking a mile a minute with both excitement and dread, but the federal archaeologist immediately understood the potential significance of this artifact and issued Lee an emergency permit. When Lee got the permit, he returned to the site to make sure he hadn't left anything of significance behind. "Well, the next day it snowed," he said, and the snow covered that part of the ice patch. "It hasn't been exposed since." He had gotten the artifact just in time.

The branch was discovered to be the foreshaft of an atlatl made from a birch-bark sapling. It was 10,300 years old, and to this day, the atlatl foreshaft is the oldest known ice-patch artifact in the world. On closer inspection, Lee noticed three evenly spaced notches on either side of the weapon. "Those markings were probably made by the hunter to indicate ownership," he told me. The hunter had expected to get his weapon back.

Lee recounted the tale of this discovery to a small group of archaeologists, climate scientists, and one writer while we ate lunch at a rocky outcrop just above the ice patch. Far below, a sapphire-blue lake shimmered in the early autumn sun. We were perched at 10,500 feet above

sea level. Cold gusts of wind swept up the mountain, so we tucked ourselves against a low rock wall while Lee spoke. Even then, fifteen years
later, I could hear the excitement in his voice.

After lunch we all wandered down to the ice patch to explore on our
own. The ice patch was large—seven hundred feet long and four hundred feet wide. Its roughly six acres of ice lay in a hollow along a steep
slope. I slid across its slushy surface and down to the bottom edge. As
I traversed the forefield of the ice patch, I hoped to find something
spectacular, too, like the remaining pieces of the atlatl or perhaps an
arrowhead knapped from an obsidian core, but I only found a slurry of
gravel, mud, and meltwater. I never did get the eye for finding artifacts,
though it wasn't for lack of trying.

Coring the Ice

Aside from preserving artifacts, ice patches open a new window into
past climate history for mid-latitude regions like the Rocky Mountains
and the Alps. Usually, climate records for these regions are only found
in lake sediment cores, trees, and mineral formations found in caves.
But these records often have large gaps or are limited to recent centuries. Ice cores from stable ice patches, on the other hand, have the
potential for being much older. Getting them is another story.

In late summer 2016, Nathan Chellman, a post-doctoral fellow at
the Desert Research Institute in Reno, visited the ice patch from which
Lee had collected the atlatl. His goal was to collect ice cores and map
the internal structure of the ice patch using ground-penetrating radar
(GPR). GPR works by sending and receiving pulses of electromagnetic
energy through a medium—for example, ice. Glaciologists also use
GPR to measure the thickness of the ice, map the layers in the ice, and
see the underlying topography between the glacier and the bedrock. It
works like sonar. A signal is sent into the ice, and when it encounters

an object, the signal bounces back to the receiver while other signals continue to penetrate the ice. The data are recorded as the time it takes for the energy to penetrate the ice and bounce back to the receiver. The time is then converted to depth.

Choosing the right frequency is key. Higher frequencies allow for more details in the ice to emerge, but the pulses of energy can't penetrate as deep, which means that there is a chance that the signal wouldn't reach the bottom of the ice patch. Low frequencies, on the other hand, penetrate the ice to greater depths than higher frequencies, but you wind up with an overview of what the ice patch looks like and not much detail. It's a tradeoff between being able to map the entire ice patch down to the bottom versus seeing individual layers in the ice.

I asked Chellman if you could detect artifacts with GPR technology. "The artifact would have to be really large, like three feet, to see," he replied. Archaeologists do use GPR to detect features below the ground like structures and burials, but so far, no ice-patch artifacts have been detected with GPR.

To collect data on the ice patch, Chellman put the GPR device on a teal plastic sled, securely attached with rope. From the top of the ice patch, he then lowered the sled with the GPR device—the same sled on which he used to career down hills as a child—onto the ice patch and let it slowly slide down the patch until it reached the bottom edge. The GPR weighs about forty pounds, so it wasn't easy to get it to slide smoothly down the ice. He then dragged it back up and moved several feet over to do it again. He repeated this until he had mapped the entire ice patch.

Since he wouldn't be able to process the GPR data until he was back at the lab in Reno, he had to guess where the deepest (and hence oldest) ice in the ice patch might be to extract a core. He chose a spot midway down the slope of the ice patch and just in from the edge. The device was a lightweight double-barrel electric drill powered by a portable generator. It's called the Prairie Dog.

As Chellman powered up the drill and began coring the ice patch, something he's done many times in Greenland, the drill suddenly stopped working. It was stuck. Linda Jarrett had the same experience when drilling ice-patch cores in Norway for her PhD project. She told me that the drill she used got so stuck in the ice that she was unable to get it out. She had to leave it there, and as far as she knows, it's still stuck in the ice patch. Erik Blake, who helped drill an ice core at the Thandlät ice patch in the Yukon, experienced the same thing, although he managed to finally extract his drill. In fact, everyone I spoke with who attempted to extract cores from ice patches lost or almost lost their drills to the ice. Both Jarrett and Blake ended up using saws to collect square blocks of ice, which was far less elegant than a nice round core. But it worked.

The reason ice patches pose such a challenge is that there is a lot of water in them. With glaciers, there's usually very little meltwater in the ice, especially because scientists typically drill alpine glaciers somewhere near the top in the accumulation zone, where average temperatures are below freezing. At the poles, persistently cold temperatures keep the ice frozen. Because ice patches are below the equilibrium line of glaciers, there is a lot more melting going on, especially considering the amount of snow that can accumulate atop an ice patch. As Chellman drilled into the ice patch, the meltwater froze around the metal and the drill jammed. The difficulty in drilling ice patches has to do with how they form.

Ice-patch scientists, including Chellman, think that ice patches accumulate ice through the percolation of meltwater rather than compaction of snow like glaciers. As snow melts atop an ice patch, it trickles down until it reaches the interface between the cold, impermeable existing ice surface, which is kept frozen by an underlying layer of permafrost, and the existing snow surface. Jarrett's research showed that the annual snow accumulation can be nearly thirty feet thick, perhaps thicker. This meltwater is either refrozen early in the melt season while the ice patch retains the previous winter's cold temperatures or at the end of the season, with the onset of freezing wintertime temperatures. The key is that for ice to accumulate, there needs to be an overlying layer of snowpack that provides a source of meltwater and a layer of insulation. Because the meltwater percolates through the snow, it ends up frozen as a clear ice layer. As the snow melts down, any organic material ends up sitting on top of the clear layer of ice.

Eventually, Chellman managed to get the drill free and extract three cores from the ice patch. The longest and most intact core measured more than eighteen feet long. In the field, he separated the clear ice from the organic dark layers. There were twenty-nine organic layers bounded by clear ice. He and his colleagues sent the organic layers to Woods Hole Oceanographic Institution for radiocarbon dating and took the clean ice layers back to the Desert Research Institute for isotope analysis.

Ice-Core Sampling

To find out how ice cores are processed, I visited Chellman at the Desert Research Institute one spring day. When I arrived at the lab on the second floor of DRI, Chellman was focused on one of eleven computer monitors lining a long wall. Multicolored wavy lines slid by on the screen. On the opposite wall was a dizzying array of wires, tubes, and instruments that measure the concentration of elements found in the ice.

Although Joe McConnell, the Ice Core Laboratory's director of research, calls it "spaghetti," it's actually an impressive state-of-the-art system that McConnell designed himself. McConnell and Chellman spend much of their time studying human impacts on remote regions of Antarctica and Greenland by looking for trace elements trapped in the ice.

Chellman greeted me with his usual good cheer, and after a brief bit of banter over the fact that there isn't a hotel in Reno without a casino (I was staying at the Atlantis, where my husband was busy trying his luck at the slots), he offered to show me around the lab starting with the "cold" room, which is where the ice cores are processed. The room's temperature is kept at a chilly -4°F (-20°C) to keep the ice cores from melting while they are being handled. Before entering the cold room, Chellman handed me an oversize red parka, like the kind worn by modern Arctic explorers, and a pair of equally oversize fleece gloves over which I pulled on a pair of blue latex gloves. The cold room was just large enough for two people to work together processing the ice. For now, I observed Chellman as he explained how the cores are prepped for analysis.

Although it was only ten o'clock in the morning, Chellman had been in the lab for four hours getting the system ready. I had arrived on the second of three days he'd dedicated to running each of the dozens of forty-eight-inch-long cores through instruments so sensitive that they can detect trace elements on the order of parts per quadrillion. Parts per quadrillion is the number one followed by fifteen zeros. One part per quadrillion would be like searching for something the size of a dollar bill in all of Canada.

The entire length of ice Chellman planned to process measured five hundred feet. It came from an almost ten-thousand-foot-long core drilled at the center of Greenland's ice sheet in the early 1990s as part of the United States' Greenland Ice Sheet Project-2, or GISP2. At the time, the GISP2 core was the longest core drilled anywhere in the world. It took the drilling team five seasons to reach bedrock.

The ice collected during the GISP2 project has been stored in the National Science Foundation Ice Core Facility in Lakewood, Colorado. Row upon row of silver tubes house almost fourteen miles of ice from drilling projects in Greenland, Antarctica, and North America. The storage room is roughly two thirds the size of an Olympic pool. The temperature is held at a steady -33°F (-36°C). Scientists like Chellman can submit proposals to access this ice, but it's very competitive because the ice is destroyed during analysis. "You only get one shot at getting the data you need," Chellman tells me. Once the ice is gone, it's gone. Archives like the Ice Core Facility are irreplaceable.

The section of the GISP2 core that Chellman had applied for was between thirty-one thousand and forty-one thousand years old. He planned to compare data from this core to more recent ice cores that span the last two thousand years. He was interested in teasing apart two different sources of black carbon—the first from wildfires and the second from human activities, such as carbon produced during the industrial revolution, and then linking these to climate variability in the past. Due to prevailing winds, ice cores from the Arctic and Greenland record events that occurred in the northern hemisphere while ice from Antarctica records events from the southern hemisphere. Events that occur near the equator may be recorded at both poles.

Before Chellman fed the ice into a deceptively simple-looking contraption, he shaved a thin layer off the outer core. Although wrapped in plastic, ice cores are transported in insulated cardboard boxes, which helps keep the ice as cold as possible as it travels from wherever it was collected to Lakewood. Shaving the core removes any debris from the packing material that could skew the results.

Once he finished shaving the core, he used a bandsaw to remove a thin outer layer and get to the inner core of the ice, slicing a tiny fraction off each end of the core, but not too much since every sliver removed represents time—a slice of lost data. When drilling deep cores, the pressure of the ice at depth is so great that the hole wants to close in on itself. To keep that from happening, the drill hole is filled with a fluid that counteracts that pressure and keeps the hole open. In the past, diesel was the fluid of choice. "They use different chemicals now, but they are still pretty gross," says Chellman. Trimming the core removes any potential contaminants so that when the ice is processed, only pure ice is sampled. Ice patches aren't deep enough to require a drilling fluid.

"You really need two people to run the system," said Chellman. "One person to process the ice in the cold room and another to watch the monitors in the lab to make sure the equipment is running smoothly." He then placed the trimmed core in an open tube that held it in place above a custom-designed heated plate. The tube feeds the core onto the heated plate, where the ice melts. The heated plate separates the meltwater into three parts. The outer layer is used to look at pollen and other plant materials (or other substances that are difficult to contaminate, since the outer layer is most prone to contamination) trapped in the ice. An inner layer gets fed to the lab, where instruments read oxygen and hydrogen isotopes for temperature proxies. And the innermost, and therefore cleanest, ice goes to the "clean" room, where a suite of high-precision instruments measures trace elements. No one goes into the clean room without donning a sterilized white bunny suit.

Out in the lab, computer monitors chart the ups and downs of various chemical constituents and isotopes present in the ice. And by looking at the isotopes, Chellman can reconstruct the climate history. Every now and again the sulfur line would spike, indicating a volcanic eruption from long ago. Hundreds of years of history pass by in minutes.

Accurately dating the ice is key to interpreting the results on the computer monitors. There are several ways to date glacial ice. Although dating the ice using oxygen, uranium, and beryllium isotopes is possible, it's much easier to determine age by counting the layers. High winter snowfall creates annual layers in the ice. These winter accumulation layers are bounded by summer melt layers. Glaciologists can tell these layers apart and count them like tree rings—at least to a point. As the depth of the ice increases, the rings become so compressed that counting them becomes impossible.

To get around this problem, glaciologists can accurately constrain the dates of an ice core by matching the signals from major events of known dates. These events are called "tie points," and they are usually major volcanic eruptions that have deposited sulfuric horizons that are visible in the ice. These dating methods work well for polar glaciers and ice sheets because they move much more slowly over flatter surfaces than non-polar glaciers, so they don't deform as much at their bases.

Later that day, Chellman pulled up the results of the GPR measurements from the ice patch he cored near Yellowstone. A figure with lots of squiggly lines all layered on top of one another appeared on the screen. The GPR had mapped the entire ice patch all the way down to the bottom. I could even see the different layers of ice. The GPR data had successfully captured all of the layers of ice that Chellman saw in the ice core.

The organic layers he sent to Woods Hole Oceanographic Institution matched the age of the atlatl. The ice was 10,300 years old. With perennial ice patches, organic layers constrain the dates of the ice much like volcanic eruptions create tie points for dating glaciers. Radiocarbon dating of the various layers suggested that the ice patch gained ice throughout the last ten thousand years, even during warmer periods when there was less snow thanks to drought or warmer periods.

The ice patch had survived multiple warm periods over the last ten thousand years, including a four-hundred-year span between 4200 and 3800 years ago. However, the intense warming in the Greater Yellowstone ecosystem since 2000 is unprecedented. Temperatures surpassed the Medieval Warm Period from 950 to 1250 CE, which wasn't a centuries-long heat wave, but rather was characterized by warmer summer temperatures with lots of variability over decades. The same is true of the Little Ice Age. This is fundamentally different from current climate change, which is much more intense and rapid. Warmer temperatures mean that water in the atmosphere tends to fall as rain rather than snow. But it's not just how much snow there is, but how quickly and when the snowpack melts.

A 500-Year Flood

On June 10, 2022, in Gardiner, Montana, at the north entrance for Yellowstone National Park, a light rain began to fall. The rain grew heavier as the day matured and continued to fall for the next three days.

The National Weather Service issued flood advisories and warnings throughout Wyoming, Montana, and Idaho. That it was raining wasn't so much the issue, but the fact that it was rain falling atop the six feet of snow dumped in the mountains over Memorial Day weekend made everyone worry. Yellowstone managers proactively closed several roads in the park on June 12 just in case. The rivers were running unusually high. Snow monitoring sites in the mountains were off the charts for that time of year.

By six o'clock the next morning, the Yellowstone River just north of Gardiner had risen so high that parts of Highway 89, the only road into the northern end of the park, were covered by water. State officials closed the road. The road from Mammoth in Yellowstone to Gardiner along the Gardner River also closed. Everyone in Gardiner, residents and visitors alike, was trapped. Two hours after the road closure, a bridge crossing the Yellowstone River collapsed. In the small community of Silver Gate at the park's northeast entrance, river water from Soda Butte Creek spilled over its banks and onto the road, washing out a section of pavement.

Yellowstone headquarters at Mammoth Hot Springs lost power, and then the sewage line that carries wastewater from Mammoth down to the treatment plant in Gardiner burst. Tens of thousands of gallons of wastewater spilled into the otherwise pristine Yellowstone River. The road washed out in three places, and by midday the Yellowstone River undercut a bank, and a multi-family home simply fell into the river and floated away. Fortunately, no one was inside, but those who lived there lost everything they owned. In the weeks following the flood, their belongings wound up miles downstream.

The flood might never have occurred had it not snowed so much in the mountains two weeks earlier. You could say it was bad luck, but more variable weather is the new normal, and we can't begin to anticipate back-to-back climatic extremes like these. Although climate scientists know that the frequency of extreme weather events has increased, predicting when they will occur remains elusive.

Park Superintendent Cam Sholly closed the park to all visitors and evacuated more than ten thousand park visitors. Everyone in the communities of Silver Gate and Cooke City on the northeastern edge of the park was stranded. Helicopters ran nonstop to evacuate everyone. Miraculously, no one was killed or seriously injured. Had roads not been closed early, the outcome surely would have been different.

Around eight inches of water entered streams and rivers in just twenty-four hours. Much of the water came not from rain, but from melting snow from the winter storm ten days earlier. A stream gauge in the Gardner River recorded a record-breaking flow of 2890 cubic feet per second (CFS), which was four times the average peak flow. In the Yellowstone River, the peak flow measured 49,400 CFS, whereas it's normally around 12,000 CFS. Rates of discharge shattered the previous record set in 1918.

But this flood was different. The flood in June 2022 was classified as a 500-year flood event, which means that there was a one in five hundred chance of a flood of that magnitude occurring in any given year. Put another way, there was a 0.2 percent chance of an event like that happening. But this doesn't mean that an event like this won't occur for another 500 years. In fact, events like these are becoming more common because extremes in temperature and precipitation are more frequent. Even as I write these words exactly one year after the flood, Yellowstone National Park is considered to be in moderate to severe drought, as it had been the year of the flood. In fact, 2022 was the fourth driest year in recorded history. Most of the park has been in drought conditions for the last several years.

Montana is known for extreme shifts in weather. It can and often does snow even in summer. But with the flood, the state experienced the most significant catastrophe in Montana's recorded history and certainly in the minds of residents. It's not just the amount of snow that falls in the mountains; it's equally important how fast the snow comes off the mountains during spring and early summer. The most extreme river flows have not only increased in frequency but have increased in magnitude as well. The timing has also shifted toward earlier-season peak flows. If we allow our baselines to constantly shift, who's to say what's normal?

Environmental Generational Amnesia

I arrived in Bozeman, Montana, in late spring almost twenty years ago. The lawns of Montana State University campus were mowed into a tidy green crosshatch, and I remember flicking off my shoes and having lunch under a blossoming crabapple tree. The sun-warmed grass was soft under my feet. It was my first spring in the Rocky Mountains, and I did not yet understand that winter had only retreated a few weeks before and that it could return at any time.

It took me four days to drive across the northern states, starting out on Highway 84 West in Connecticut. I drove over the Hudson River into the southern sliver of New York. I hardly had time to take New York in before I crossed into Pennsylvania—a deceptively large eastern state whose width can only be appreciated on a solo journey. I would not have known the difference between these artificial boundaries if it weren't for the green or blue state highway signs welcoming me across each border. The hardwood forests stretched on regardless. And although thicker than they were a century ago, the eastern hardwood forests are still youthful. Walking through them, you can find the remains of stone walls that once bordered open fields and crumbling foundations with caved-in hearths partially hidden by maples and birch. It will be a century or longer before beech and hemlock reclaim these forests and they become old again, although there are pockets tucked away that were never cleared.

The landscape sprang open when I reached the prairies of Minnesota. There were no trees save for those planted purposefully around homesteads. It seemed strange to me to see trees so obviously placed in an otherwise treeless landscape, but then the wind would be strong across the prairie and the homestead would need a windbreak. South Dakota was altogether different—dry, stark, limitless. I drove through the badlands and wished I had made the time to explore their story told in layers of rock.

I can't remember when I first saw the Rockies, probably somewhere in eastern Montana. Their broken tops pierced the bright blue sky. Other than the busyness of Billings, I didn't drive through any towns of appreciable size until I arrived in Bozeman. Still, at a population of about forty-five thousand people at the time, it was small by comparison to the cities of my eastern upbringing. I was told to watch out for "that 19th street." I happened to be on 19th street at the time, and as I looked out across the four lanes of traffic, two in either direction, I wondered what all the fuss was about. But my guide had obviously seen something I hadn't—a time before there were four lanes of traffic, or at least a time before there was so much traffic. He wasn't an ancient old-timer of Bozeman, just someone who had seen a shift imperceptible to those without memory of anything else.

Change sometimes happens all at once like an unexpected thunderstorm erupting overhead, or a flood severe in its abruptness. Other changes happen over millennia, like the erosion of the Appalachian Mountains to mere hills. But somewhere in between are the changes in the environment that occur over generations, perceptible by those still living but forgotten by those who come after. Or maybe not forgotten so much as erased from our collective memory. There's a term for this— it's called environmental generational amnesia. Coined by psychologist Peter H. Kahn Jr., it refers to how each generation perceives environmental degradation based on the condition of their environment as they experienced it in childhood. Our childhood serves as our baseline; the problem is that our baseline is often already degraded.

In a series of studies, Kahn asked children growing up in Houston about the environment they lived in. Most agreed that air pollution was a bad thing, but when asked about whether they thought the air pollution was bad in their own city, many of the children did not believe they lived in a polluted environment, even though Houston remains one of the most polluted cities in the United States. The polluted city of their childhood will serve as the baseline to which they compare Houston's pollution in the future.

In his book *The End of Night*, Paul Bogard writes about the loss of dark night skies throughout the world—not just the developed world, but almost everywhere. Sky glow from cities scatters and penetrates even the most remote places on Earth. We have lit up the night so that we don't have to face the beast in the shadows. But over time, what began with gas lamps has become gas stations lit so brightly that it hurts to look at them. Bright lights such as these, and even the ones we put up to light our driveways, cast so far and wide that it is Earth seen from a distance that looks like the night sky. It's hard to believe, but there are children who have never seen a dark night sky and would be afraid to sit in darkness lit only by starlight. It is even harder to believe there are adults who've never seen a starry night except the one painted by van Gogh. This is how our baseline shifts.

Even within a lifetime, we forget. I am alarmed at how many of the details of my own life I've forgotten. I can't remember the name of my third-grade teacher, nor do I have any concrete memories of my great grandfather, only a still-frame picture in my mind. This is why I write, to remember plucking periwinkles off rocks in tidal pools on a Maine beach, then holding them to the rock again until they reattach. To remember walking the Marginal Way along Maine's southern coast with my parents, sister, and grandparents, then getting chocolate-and-vanilla swirl soft-serve ice cream in a cake cone at the end. Because there is something deep within myself that has been forgotten and is dying to be remembered. Maybe we all share this feeling.

After a brief stay in Bozeman, I drove the seventy-five miles south to Yellowstone National Park, and I've never left—at least not permanently. Yellowstone became a national park in 1872—the first of its kind and predating the National Park Service by forty-four years. In some ways it feels like I am being given the chance to remember something that has been forgotten. I have no memory of Bozeman from one hundred years ago. But by comparison to Paradise Valley, which is mostly open agricultural land that is slowly being invaded by ranchettes, and by comparison to Yellowstone, which is much like it was one or two hundred years ago, I begin to reach back in time through space. At times I have felt rather isolated in Yellowstone, blocked off from the environmental catastrophes of the rest of the world. But that was before the flood and before the wildfires. I live in a protected area, and although I'm lucky to do so, it's important for me to be reminded that this place does not stand alone. It is part of a larger landscape that is connected to and influenced by not only the towns, roads, and rural development outside the park, but also global air quality and temperature.

Edward O. Wilson coined the term biophilia in 1984, which he defines as the innate need for each one of us to connect with the natural world. He argues that this desire is part of our genetic makeup and is therefore inescapable. By visiting as wild a place as we can find, and as often as we can find it, we may begin to rebuild our collective memory and recover from our amnesia. How can we be good ancestors for future generations?

In the 2023 legal case of *Held v. Montana*, sixteen kids from ages five to twenty-two sued the state to protect their rights to a healthy environment for now and the future. They fought against the state's undying support of the fossil-fuel industry and fought for the transition to clean energy by 2050. Among dozens of similar cases filed across the United States in the last decade, *Held v. Montana* was the first to reach trial. In August 2023, a judge sided with the sixteen young plaintiffs, but it's up to Montana's legislature to figure out how to comply with the decision. And it's up to Montana's voters to make sure they comply.

The Montana state constitution guarantees the right to a healthy environment. It's one of the few states with this kind of language written into its constitution. Notably, the defense did not argue that climate change is a myth. That argument has been so thoroughly debunked that lawyers had to come up with a different one. The defense argued, weakly, that Montana can't be held responsible for global climate change and that Montana's contribution is negligible. But then who should be held responsible? Aren't we all responsible? What contribution to global climate change is acceptable? Perhaps by reorienting our relationship to the environment, we will become more aware of our individual contributions and responsibilities to each other and the ecosystem of which we are a part.

Modern western society tends to view mountains for the resources they provide—fresh water, minerals, timber, hydropower, and other tangible and monetary benefits. These are all important ways mountains nourish us, but they fail to acknowledge the immaterial aspects of mountains, which have been the object of aesthetic contemplation for millennia. Mountains are sources of artistic inspiration. They are where we ski in winter and hike in summer. Mountains are also sacred landscapes, and they have been looked upon as sacred for far longer than they have been seen as "things." Nowhere is this more evident than in Peru, where tens of thousands of pilgrims embark on an ancient pilgrimage to a glacier high in the Andes.

NINE

Pilgrimage to the Andes

"Gringos!" The Quechua Elder shouted from his perch on a boulder alongside the dustiest trail I've ever walked. I looked up and laughed with him. His eyes were bright and cheery, and he had rounded flushed cheeks.

"Si, Gringos," I replied through a mouthful of trail-dusted teeth. I laughed with him a bit less enthusiastically than I might have otherwise, but not because I wasn't amused by his observation. It was just that at 13,320 feet, after a large Peruvian lunch, and exposed to the midday sun of the central Andes, I was out of breath.

Three hours earlier our small group of mostly gringo pilgrims (five of us including my husband John), our guide Arturo Mantilla Quispe, cooks, and porters had arrived in the rural Peruvian town of Mahuay-ani, at the foot of the Sinakara Valley eighty-one miles east of Cusco. If we were to have continued east over the crest of the Andes, we'd have reached the rainforest and the headwaters of the Amazon River. But we weren't going to the jungle. We stayed high in the mountains where the air was so dry, my lips began to crack a few hours after arriving. It was the early summer dry season, although Quispe told us that during the rainy season, which was still many months away, these puna grass-lands erupt in lively shades of green.

I had come to witness Qoyllur Rit'i—a three-day pilgrimage and cele-bration of the stars, glaciers, and mountains. Qoyllur Rit'i (pronounced coil-yure-ree-tee) means "snow star" in Quechua. Quechua is a family of languages spoken by more than eight million Indigenous people of the central Andes, most of whom live in what is now the Peruvian highlands. Qoyllur Rit'i was named after the first rising star of the night, which is actually the planet Venus. Venus appears to briefly rest atop the Qolqepunku Glacier at the head of the Sinakara Valley before continuing its ascent in the night sky.

The Conquest of the Inca

Quechua speakers include numerous diverse ethnic groups like the Chanka of Ayacucho in Peru, the Cañari of Tumebamba in Ecuador, and the Qulla of western Bolivia. Some words we commonly use today, such as puma, condor, llama, and coca, are Quechua words that have been assimilated into the English language. Quechua was also the language spoken by the Incas. While many people see the Quechua as direct descendants of the Inca, that's not quite true. Inca means "king" or "ruler" in Quechua. What we have come to know as the Inca were simply the last in a long line of pre-Columbian civilizations that rose to power before the Spanish conquest.

The Inca empire was one of the largest and most powerful empires in the pre-Columbian Americas, yet they only ruled for about a hundred years. The Inca adopted Quechua as their official language after conquering Cusco in the 1430s. As the Inca empire grew, the Quechua language spread. At its peak, the Inca empire stretched from present-day Ecuador in the north to Chile and Argentina in the south, covering an area of more than 770,000 square miles.

The Inca referred to themselves as Tawantinsuyu, which means "land of the four quarters" or "four united regions." The four regions included Chinchaysuyu to the northwest, Antisuyu in the northeast, Kuntisuyu in the southwest, and Qullasuyu in the southeast. The four regions met at the Inca capital of Cusco. To mark the boundaries of their empire and hold important events, the Inca built shrines atop more than a hundred of the most prominent peaks. Before the Inca rose to power, mountains were worshipped at a distance.

In 1531, after years of skirmishes and exploratory missions by the Spanish looking to expand their own empire in the Americas, the Inca empire began to crumble when Spanish conquistador Francisco Pizarro arrived with his small army of 168 soldiers, one cannon, and twenty-seven horses. Two years later the Spanish had taken Cusco. Following the Spanish conquest, a few members of the Inca royal family established the small, independent Neo-Inca State in Vilca-bamba, which was located in the relatively inaccessible Upper Amazon to the northeast of Cusco. But even this last stronghold fell in 1572. Although many died of direct warfare and displacement, diseases like smallpox and measles killed far more people, which helped solidify Spanish control.

The Spanish set about attempting to destroy Indigenous culture by banning Indigenous religion, building Catholic churches, walling over important temples with Spanish architecture, and destroying sacred relics. In the days before traveling to Mahuayani, my husband John and I explored Cusco, a vibrant city full of stunning architecture that reveals the city's tumultuous past. While the massive Spanish cathedrals are the most prominent structures in the city, the finely worked stone of the Inca often peeks out of the Spanish facades. Yucay limestone and black andesite blocks were so perfectly shaped that they had no need for mortar. Once I caught an eye for spotting them, the imprints of the Inca, and the earlier Wari culture, were everywhere.

But the Spanish couldn't or were unwilling to ascend the highest peaks where mountaintop shrines and human sacrificial altars had been built, or perhaps they were unaware of these mountaintop shrines. During the short time of Inca rule, they built shrines atop nearly two hundred snow-capped peaks at elevations over sixteen thousand feet. The Inca believed that the mountains were the protectors of their empire and that they provided strength and stability to the people. Mountains were the gatekeepers to the spiritual realm, and they held the power to connect the living with the dead. They believed that by climbing the mountains, they could reach the heavens and communicate with the gods.

While there is no mountaintop shrine on Mount Sinakara, in Quechua cosmology, every snow-capped mountain possesses its own deity known as an apu. Mount Sinakara's apu is El Señor de Qoyllur Rit'i— God of the Snow Star.

The Origin of Qoyllur Rit'i

During the three-day festival, as many as one hundred thousand pilgrims, representing Indigenous groups from the mountains to the rainforest, come to pray for the health of their animals and for a good harvest in the coming year. The year I visited (2022) was the first year the festival took place since the COVID-19 pandemic struck in early 2020. Although the number of people gathered was only a fraction of past crowds, the mood was jovial and festive.

After lunch, Quispe set us free to explore Mahuayani before beginning our pilgrimage up the valley. He told us that the rest would give us time to acclimate to the altitude, but the two hours we spent in town wasn't enough time for that. I think he wanted to see if any of us keeled over before he risked taking us up the mountain. Under a bright blue Andean sky, Mahuayani bustled with pilgrims readying themselves for the festival. John and I made our way down to the muddy brown river that flows along the edge of town, where a popup market featured vendors selling jewelry, crosses, paintings of Jesus, sunscreen, hats, alpaca scarves, and colorful blankets.

I purchased a dozen huayruro-seed bracelets for about seventy-five soles, or about fifteen dollars. The seeds are red and black, smooth, and a bit larger than a black bean. I slipped one bracelet on my wrist and gave another to John. The rest I planned to give away back home. In Quechua cosmology, red is the color of female energy, which is called Pachamama, or "Mother Earth." It represents life, menses, and birth. Black represents the immaterial and male energy of the universe. Together, the red and black huayruro seeds are said to gather positive energy and repel negative energy. Many Quechua mothers place small huayruro-seed bracelets on the wrists of newborns as protection.

Two hours later and still upright despite the thin air, we began our journey up the Sinakara Valley. The pilgrimage starts at a small mud-brick church warmed by the light of a dozen candles. Many pilgrims had left a trinity of coca leaves called a kintu as an offering. The three leaves represent the three worlds in Quechan cosmology—the living, the dead, and the supernatural. Pilgrims who seek permission to walk the path in this sacred landscape make these offerings to gain protection from the mountain spirits. But coca is also a stimulant; coca leaves are harvested from the same plant used to make cocaine. Indigenous Andean people chew the leaves to boost energy during long days working their land.

Although the pilgrimage is rooted in Indigenous Andean traditions, it is riddled with Catholic elements. The main celebration day takes place on the fifty-eighth day after Easter Sunday and the week before the feast of Corpus Christi, when the moon is full. Ethnographers call Qoyllur Rit'i a syncretic religious pilgrimage—an event that merges two worldviews—but the timing and relation to Corpus Christi was an attempt by the Catholic Church to replace Quechua beliefs rather than to merge with them.

As we continued up the trail, I asked our guide Quispe about the origins of the festival. Quispe, quiet and a bit melancholy, began by telling us about Marianito—a twelve-year-old boy who lived in the Sinakara Valley with his family more than two hundred years ago. Quispe said that Marianito's family shepherded alpacas, llamas, and rams along the valley's slopes. There are still many small alpaca farms tucked into hillsides throughout the valley. The homes are made of dust-colored stone with stone fences and stone corrals that blend in so well with their surroundings I almost failed to notice them.

On the trail, we were joined by women carrying infants in *q'ipirinas*, which are colorfully striped Indigenous Andean blankets, on their backs. Chubby-cheeked infants peered out from the blankets' folds. Others used q'ipirinas to carry bundles of wares they planned to sell on the mountain. Those with especially large loads and the money to hire them opted for mules. Traditional Andean clothing is vibrant, especially clothing worn by women. Colorful wide-brimmed hats shade them from the intensity of the sun. Their embroidered skirts bloom wide at their knees, perfect for kneeling while planting potatoes. The colors and patterns are more than decorative; skirts, called polleras, are trimmed with patterns specific to certain regions. For those in the know, the designs of the polleras communicate where their wearers are from.

Quispe continued. Marianito felt mistreated by his older brother like many younger siblings do. One day Marianito had had enough, and he took the few animals his father had entrusted him with high up the Sinakara Valley near the foot of the Qolqepunku Glacier. There he met a blond, fair-skinned boy about his age who called himself Manuel. The fair-skinned boy comforted Marianito and gave him bread. The two boys played and danced together for several days until another herder in the valley spotted them. The concerned neighbor tattled on them.

When Marianito's father learned of his son's friendship with a stranger, he went to look for him. When he reached the foot of the glacier, he saw that not only had his herd grown fat and woolly, but there were more of them. Marianito's father was astonished by the transformation of the herd and decided to reward the boys by giving them new clothes. (There are a couple of different versions of what follows, but this is what Quispe told us.)

Marianito cut a fragment of Manuel's garment and took it to Cusco so a tailor could supply new material, but the material was so delicate and unusual that the tailor could not help Marianito.

When the Bishop of Cusco learned of this fabric, which was the same textile used to sew the bishop's robe, he suspected the boy of thievery. The bishop sent the parson of a nearby village to investigate. The priest, sacristan, church steward, and Marianito climbed into the mountains to find Manuel, only to discover him dressed in a beautiful white robe. Marianito ran toward his friend. At that moment a bright white light flashed, temporarily blinding the parish priest and his companions. Suspecting trickery, the priest left and came back a few days later with local authorities. But when one of them tried to catch Manuel, he was left holding a tayanka bush in the shape of a cross. Marianito, frightened at the transformation of his friend, died suddenly near a big black rock. In mourning, his family buried him there.

As word spread of the unusual events in the Sinakara Valley, people came to light candles and pray near Marianito's grave. To avoid a cult of the black rock, the Church of Ocongate—the local district church—commissioned a Cusco artist to paint the crucified Christ on the rock in 1935. It is this image that is known today as Señor de Qoyllur Rit'i. In 1944, the image was blessed by the archbishop of Cusco. Many pilgrims believe that the spirit of Manuel still lives in the rock and surrounding glacial landscape.

The Sinakara Valley is long and narrow. The trail to its head is only about five miles, yet it took us three or four hours to walk it. Along the trail, we passed large crosses wrapped in colorful sashes. There are nine of them between Mahuayani and the head of the valley. "Each one faces a different mountain or apu," said Quispe. Pilgrims stop at each one to pray and make an offering. The low rock walls surrounding each cross were black from burning candles.

The Catholic-inspired story of Marianito and Manuel was invented soon after the culmination of a three-year uprising of Quechua-speaking rebels against Spanish colonial rule. The rebellion was led by Túpac Amaru II over forced labor, high taxation, and cultural oppression. Born José Gabriel Condorcanqui Noguera, he changed his name to Túpac, claiming to be a descendant of the last Neo-Inca king of the same name. The rebellion lasted three years, from 1780 to 1783, though Túpac and his family were brutally executed during the first few months of the revolt. His twelve-year-old son was spared but spent the rest of his life in prison.

The revolt ended in one hundred thousand Indigenous deaths. It also led to the banning of the Quechua language and the abolition of Indigenous clothing, which was unenforceable and largely forgotten. It is out of this uprising that the Catholic elements of Qoyllur Rit'i were introduced. The bishop featured in the legend of Marianito and Manuel did not want to lose control over his territory, which was a rebel stronghold during the rebellion. To maintain control, Bishop Juan Manuel Moscoso y Peralta reinterpreted the story in Christian terms, and Qoyllur Rit'i as it is practiced today was born.

The True Origin of Qoyllur Rit'i

Despite the Catholic church's attempt to erase Indigenous Andean traditions, Qoyllur Rit'i remains rooted in Quechuan cosmology. The true origin of Qoyllur Rit'i lies in the stars. Traditionally, Qoyllur Rit'i is a celebration of the return of the Pleiades constellation in the southern sky. The Pleiades constellation, also known as the Seven Sisters, disappears from the southern sky in late April. In Quechua cosmology, when the Pleiades constellation is not visible, the world becomes a dangerous place—one of imbalance and illness. Along with the spiritual aspects attributed to the Pleiades cluster, the reappearance of this constellation in early summer has a practical purpose for local farmers.

The clarity of the Pleiades star cluster conveys information about when to plant and the quality of the upcoming harvest. The Quechua are masters of agriculture. Over millennia, they've taken the few native potatoes that grow in the Andean highlands and have cultivated thousands of new varieties. The rapid and abrupt changes in elevation from the valleys to the peaks result in elevational bands with unique micro-climates. Each variety of potato has been cultivated to grow within one of these narrow bands of temperature and rainfall.

Apart from potatoes, Quechua farmers grow other root vegetables such as olluco, mashua, and oca in addition to beans, maize, quinoa, wheat, and tarwi, which is a small bean produced by the lupine—a bluish-purple flower whose cousins I recognize from my own backyard, the Rocky Mountains.

If the sky is clear and the constellation vivid, the rains will arrive on time (typically in mid-to-late autumn) and planting occurs as usual with an expectation of a good harvest. A dim and fuzzy-looking star cluster indicates that the rains will be late or diminished, so planting is delayed and the harvest is expected to be poor.

By paying attention to the Pleiades constellation, Quechua farmers accurately predict El Niño years. During a typical El Niño year in the Andean highlands, less rain falls, and temperatures are warmer than usual. High, thin cirrus clouds develop in the atmosphere over the Andean highlands to the northeast, which is the direction the Pleia-des constellation reappears in the southern sky after its hiatus. These clouds are so thin and occur so high in the atmosphere that they aren't visible to the naked eye but contain enough water vapor to cause an apparent dimming of the Pleiades cluster.

"Ah, there is our camp," said Quispe. I looked up and saw our horses and four North Face expedition-style tents in a flat area alongside the creek at more than fifteen thousand feet. We arrived just as the sun slid behind the mountains, making it instantly cold. My head pounded with dehydration and a mild case of altitude sickness. I crawled inside the tent to rest. As soon as I slipped into the down nylon sleeping bag, waves of nausea washed over me. I skipped dinner and tossed fitfully throughout the bitterly cold night.

Earlier in the day, I had marveled at how many pilgrims passed us carrying heavy loads, seemingly unfazed by the altitude. Many of them return year after year and are used to the climb, but they are also genetically adapted for high-elevation living. The first people to arrive in the Andes probably suffered from mountain sickness, but over time their red blood cells evolved to carry more oxygen. I would have to stay up high for weeks waiting for my body to make more red blood cells to see the same benefit.

Ukukus—The Peacemakers of Qoyllur Rit'i

After a fitful night's sleep, I headed to the breakfast tent wearing everything I had slept in—wool underwear topped with a down coat and down pants, a wool hat, and thick gloves. The temperature had dropped to -10°F (-23°C) overnight. I never quite got warm. I wished Quispe a good morning and he generously handed me a steaming mug of hot coca tea. As I sipped, the sun cast its light across the tent, so I took my tea outside to soak in its warmth. The sun's effect was immediate. I turned to face it and let its warmth replace the chill of the previous night. All around, glacier finches foraged for seeds among the shrubs and grasses. Other than Emperor penguins, this is the only bird that nests directly on ice. Above me black and white mountain caracaras soared on unseen wind. After the brutally cold night, the sun had never felt so good. As the sun warmed my face, I understood why the Inca worshipped this star as a god.

After breakfast, Quispe, my husband, and I made our way up the trail to the Alacitas market, which sat beneath a blue haze of smoke from small dung-fueled warming fires (there are no trees nearby) and burning incense. At the Alacitas market, one can purchase miniatures of just about anything a person might desire. A house. A car. A baby. Faux diplomas and certificates. The belief is that on your fourth pilgrimage, El Señor de Qoyllur Rit'i will make the wishes of the devout come true.

Quispe told us that when he first made the pilgrimage twelve years ago, he bought a replica guide certificate in the hopes that one day he'd become a professional guide. "And look, here I am!" he said. But his certificate was earned with years of hard work. Aspiring guides spend three to five years training to earn their guide certificates in Peru. "But many people now come and wish for a car or a house and don't do anything to make it happen. You can't just wish for something; you need to work to make it happen," Quispe explained.

We continued past the market and crested a small hill. The main church where Marianito is said to be buried under the black rock came into view. The church was once a simple, small adobe building but it is now a sizeable stucco structure with a red roof. A large dance arena had been built along the short end of the church. It even had power. There was a line of pilgrims waiting to go inside one of the side doors to pray at the black rock. We joined them and awaited our turn.

Each of eight Quechua nations from the Cusco region sends a dance troupe to the festival; they arrive wearing ornate regalia that is specific to their cultural group. Each nation is led by a *carguyoq* (or, in Spanish, *mayordomos*) who organizes the dance troupe during the pilgrimage and carries a cross with the image of the Lord of Qoyllur Rit'i to the church. The dance troupes were just arriving along the trail below.

When I finally entered the church, it took me a moment to find what I was looking for as my eyes adjusted to the dim light. Behind a glass wall sat the black rock with the painted figure of Jesus. I paused a bit too long, and an ukuku inside the church hurried me along in a high-pitched falsetto. He was dressed in a fringed robe of black, red, and white. It worked, and I hurried out the back door of the church and onto the dance arena, which for the moment was empty. The massive Qolqepunku Glacier clinging to the gunmetal gray rock and scree came into view as my eyes readjusted to the bright light of day.

Ukukus are mediators between the spirit world and humans. They are liminal beings—somewhere between a bear and a human. They are tricksters, threshold beings of dawn and dusk. Each dance troupe includes at least one ukuku. Collectively, the ukukus are good-natured enforcers of the festival. They speak only in falsetto during Qoyllur

Rit'i. The pitch is so unusual and unexpected that most people pay attention when an ukuku speaks. And if their voice fails to get the attention of a pilgrim exhibiting bad behavior, the ukuku will tap you with his whip for things like forgetting to take off your hat before entering the church or drinking alcohol, which might garner more than a tap. Their intention isn't to harm but instead to call attention to bad behavior.

Later that evening we sat around the dinner table in the cabin tent. The cooks had set out fresh bread, tea, and hot soup. It was delicious and I wished I had more of an appetite (which had lessened due to the altitude). "Don't feel bad," said Quispe. "The cooks make the same amount of food no matter how much you eat. Eat what you can."

He told us that the cooks have been giving away all the extra food to other pilgrims and vendors at the Alacitas market. "It's a tough year for them," said Quispe. There are only about five thousand people on the pilgrimage, which is far fewer than the one hundred thousand pilgrims who usually attend. The normally large crowds are a boon to the vendors, which makes the difficult journey up the valley worth it. Even just getting to Mahuayani is a financial burden for some of the sellers. But with far fewer people, it's doubtful whether they'll make enough money for the trip to have been worth it. And I realize that this is why Quispe has been so melancholy. He said that he was upset that the pilgrimage had attracted so few people this year. He worried for the vendors. I vowed to spend the rest of my soles in the market the following day.

A few minutes later, two others in our group, Boris and Mauro, rushed into the tent. Mauro, who is from Argentina, began speaking rapidly in Spanish to Quispe. He was talking so fast and with such animation that it was obvious something was going on, but since I know only a few Spanish phrases, I waited for a translation.

Traditionally, the Quechua nations of Paucartambo and Quispicanchi embark on a twenty-four-hour pilgrimage through the mountains carrying the images of the Lord of Tayancani and the Grieving Virgin, which they brought from their temples to the temple of Qoyllur Rit'i.

They make this twenty-four-hour procession at night with only the light of the moon to guide them. It's a nineteen-mile trek with the highest point reaching 16,400 feet.

Along the way, there are five stations with crosses like those we passed on the way up the Sinakara Valley. The last station is called Intilloqsimuy, which means "rising sun" in Quechua. The station sits just above the village of Tayancani. Pilgrims kneel down next to the cross and wait through the night to greet the first rays of sun. After a ceremony, they descend to the village of Tayancani, where they leave the Lord of Tayancani in the temple, and this ends the yearly pilgrimage.

We had planned to join them the following morning. But Boris and Mauro told Quispe that the twenty-four-hour pilgrimage had been canceled. The nations were not at full capacity and felt it was unsafe to undertake such a journey with so few people. While we'd have walked it in the daylight, they'd have traveled at night and in the dark, with only the light of the moon to guide them. I was disappointed but also more than a little relieved.

Earlier in the day I had seen people walking along that path. They looked like miniatures. It's hard to get a grasp on the scale of the Andes. The mountains are so huge and so high and the glaciers so large, that it is only when a person or something for scale is in view that one can truly appreciate their grandiosity. The trail switchbacked up the hillside and then over the lip to the mountains beyond the Sinakara Valley. I desperately wanted to see what was on the other side, but I was already having a hard time at 15,092 feet. The change in plans turned out to be a blessing. By staying in the Sinakara Valley, we'd get to participate in the final day of the festival when the ukukus come down off the glacier singing, dancing, and playing drums and flutes.

As my husband and I settled into our tent for another sleepless night at altitude, the ukukus were heading to the glacier. Sometime after midnight, they had carried their nation's cross to the foot of the glacier and stayed there all night. The eight nations split up and visited the three tongues of the Qolqepunku Glacier—three at one, three at another, and two at the last. One of the glacial tongues, though, has retreated so far that the ukukus can no longer reach it safely.

Traditionally, the ukukus would carve off pieces of the glacier and bring them to Cusco. The ice is believed to possess curative properties and the ability to provide prosperity to families and land. Glacial meltwater from the mountain is the masculine energy that fertilizes Pachamama. Many pilgrims collected their own ice from the glaciers, too. The meltwater is believed to bring luck and good harvests; however, the glacier has receded so much that not only is it dangerous to approach it, but to prevent even greater loss of the glacier, it is forbidden for anyone to collect the ice.

Peru is home to 71 percent of all tropical glaciers. Since the 1970s, roughly half of the glacial ice in the Andes has disappeared. In the Andes of Chile and Peru, temperatures have risen about 0.4°F to 0.5°F (0.2°F to 0.3°C) per decade since the 1970s. Warmer temperatures have driven the permanent snow line up about 150 feet, which means that the microclimates farmers depend on for growing different potato varieties are changing.

As the Qolqepunku Glacier retreats, the bodies of ukukus who have fallen into the glacier's crevasses in previous years have begun to melt out. It's not unusual for ukukus to disappear into a crevasse. One ukuku from the Quispicanchi Nation recalled that "each year between four and five ukukus disappear in the ice." Many view the death of an ukuku as a sacrifice to the glacier. They walk the glacier at night without ropes, crampons, or ice picks. It's incredibly dangerous. The ukukus feel that the appearance of the dead is a sign of a broken relationship between humans and the sacred. One ukuku said that "the ice eats us, but then we eat it."

Agreeing to adapt to new rules around the glacier is one thing; controlling the individual response is another. Individually, most of us feel that our footprints are small and don't matter much in the scheme of the Earth. We may live seventy, eighty, even ninety years if we're lucky. Given the 4.5-billion-year history of Earth, our individual actions seem insignificant, and yet the accumulation of a lifetime of footprints matters because it is added to the accumulations of the eight billion people currently living and all those who ever lived before us.

A traditional Quechua prophecy says that the world will end when the ice turns black. Where will the apus go when the snow is gone? Where will the Indigenous farmers, who have lived in relation to the land for millennia, go?

These farmers live in a vertical world where water from the mountains is critical for growing potatoes and other root vegetables. One farmer I spoke with says that nighttime temperatures are sometimes too warm for frost to form, and frost is needed to prepare certain potato varieties for long-term storage. Farmers prepare the potatoes by stepping on them to squeeze out moisture. During the day, the dry Andean air whisks away the moisture. At night, frost helps them freeze-dry.

On the last morning of Qoyllur Rit'i, I watched as a serpentine line of pilgrims danced their way down the mountain. At first, the dancing was continuous with all nations twirling, singing, and playing music at once. It was both dissonant and harmonic. There was so much going on it was hard to know where to focus my attention. French sociologist Émile Durkheim aptly called it "collective effervescence." Like champagne that has erupted from its bottle, the mood was spirited.

Soon the atmosphere changed from chaos to order as dancers from the different nations arrived to take their turn on the dance arena. The emotion of Qoyllur Rit'i suffused the air as each nation played music, sang, and danced the story of their culture. They wore colorful, beaded panels on their backs that told cultural stories in intricate designs: Amazonian parrots for those who are from the rainforest; scenes of alpacas grazing in the puna; and farmers harvesting and planting potatoes.

After the last of the dancers entered the church, the final Catholic mass began, and the crowd fell silent. Bishops in white robes blessed pilgrims with fresh flowers dipped in holy water. As the final words of the mass were spoken, a woman next to me turned and hugged me. People everywhere turned to hug one another and wish each other well.

The crowd of five thousand was almost totally silent for a few moments before the dancing and singing resumed. Although the Qoyllur Rit'i was over, the dance troupes sang, danced, and played music all the way down the mountain back to Mahuayani.

The Children of Llullaillaco

Perhaps it's not a surprise that the highest known archaeological site in the world is located in the Andes, atop a snow-capped mountain called Llullaillaco, which is a volcano situated along the border between Argentina and Chile at more than twenty-two thousand feet. Under a platform built specifically for this purpose were the remains of three children who had been sacrificed to the gods of the mountains in Inca ceremonies called Capacocha. The dry, cold mountain air preserved these children so well, it's almost as if they had simply fallen asleep.

Archaeologist Maria Constanza Ceruti discovered the site in 1999 while serving as co-director of a National Geographic Explorer team. Ceruti is a high-altitude archaeologist and director of the Institute of High Mountain Research at the Catholic University of Salta in Argentina. She's climbed more than a hundred peaks above sixteen thousand feet in the Andes alone and hundreds of other peaks around the world in her search to understand the spiritual dimensions of mountains. Ceruti spent nearly a month on Llullaillaco surveying the area. Llullaillaco is cold and dry. The temperature averages about 9°F (-13°C) but can reach temperatures as warm as 54°F (12°C) on summer days.

In 1950, a team of mountaineers summited Llullaillaco and reported finding ruins on its slopes, but they did not find the children. During Ceruti's work, she and her team discovered three paths to the summit. The paths converged at an Inca stone structure called a tambo. Tambos were established about a day's journey apart along all Inca roads as waystations for travelers. From the tambo at 17,100 feet, a single path marked by cairns led the team up the mountain past two other archaeological sites to a height of 20,000 feet. The path ended at 21,300 feet, near a platform and a two-roomed structure right below the summit.

The Inca made substantial efforts to take their offerings to the mountain spirits as high up as possible. They seemed to prefer volcanoes and mountains that provided at least one slope that was free of ice, so that they could make it all the way to the top. Long before Saussure climbed Mont Blanc and kicked off alpinism as a sport, Inca men, women, and children climbed some of the highest peaks in the Andes as part of these Capacocha spiritual pilgrimages.

Capacocha (also spelled "Capac hucha" or "Qhapaq hucha") is a Quechua word combining *capac* (meaning "royal") and *cocha* (translated to "obligation" and sometimes "body of water"). Sacrifices occurred upon the coronation of a new ruler and upon his death, to celebrate the birth of the son of the emperor, and upon the victorious return of the Inca from battle. Natural disasters also warranted a Capacocha. Children were used as messengers to the gods because of their perceived purity and innocence. These child sacrifices were offered to the sun god Inti, to the weather or thunder deity Illapa, and to the creator Viracocha, as well as to local deities (*huacas*).

Because the Inca had no written language, much of their history is unknown, but Spanish priests chronicled some of the notable aspects of their society—at least in their eyes. Father Bernabé Cobo wrote in 1653 that "they made sacrifices to the Sun so that he would make plants grow, to the Thunder, so that he would make it rain and not hail or freeze, and to the rest of the special gods and second causes. First, they would speak with Viracocha, and afterward they would speak with the special gods. And in their sacrifices to all the universal huacas they would plead for the health of the Inca."

The Llullaillaco ceremony took place around 1500, which was around the middle of the Inca empire's reign. The sacrifices and shrines on these mountains were preceded by months of travel and processions.

Although mountain worship is found all over the world, human sacrifices are unique to the Andes.

It's horrific to think about child sacrifice. I can imagine that the children's families mourned their loss. And the children, although drugged with coca leaves and alcohol to encourage their compliance, likely felt at least some fear of their fate, even if their sacrifice was considered a great honor. The Inca sacrificed what they valued most—the pure spirits of children. Their sacrifice makes me wonder: What are we willing to do to create a healthy, livable planet? What sacrifice would be too great? While it is indeed horrific to think about child sacrifice, have we not already sacrificed our children's futures?

Windows to the Ice

Two months after returning from Qoyllur Rit'i, I accompanied a group of archaeology students and their professor on a fourteen-day adventure in the Wind River Range in west central Wyoming. We followed part of the route believed to be a long-forgotten pilgrimage trail to a glacier deep in the backcountry, probably not unlike that of Qoyllur Rit'i. The Winds, as they are locally known, are a 100-mile-long mountain chain situated along the Continental Divide. The range is named for the relentless winter gales that sweep northwest through the valley between the Absaroka and Wind River mountains.

There were seven of us, including myself. Over the fourteen-day trip, our small group had backpacked a hundred miles and gained more than five thousand feet in elevation. We had dodged lightning and spent hours in tents as it poured and thundered outside. For my part, I suffered painful blisters on both heels that rubbed raw with every step. And I loved every minute of it.

At the time of the expedition, the trip leader, Todd Guenther, taught anthropology and archaeology at Central Wyoming College—a Native

American-serving, non-tribal college in Lander (he's since retired). He is tall and slim, of Norwegian descent with a silver cowboy mustache and a western drawl to match. He bears a striking resemblance to the actor Sam Elliott.

We were in the Fitzpatrick Wilderness portion of the Winds. Wilderness, according to the US Wilderness Act of 1964, is "where the earth and its community of life is untrammeled by man, where man himself is a visitor who does not remain." To my untrained eye, there were no signs of humans. I was looking at a landscape without buildings or dams or paved roads. There were no ski lifts or mountain cabins. There was nothing we typically associate with modern human habitation, save for a well-worn trail threading through meadows and forests. There weren't even that many people. Although the land beneath our feet was now designated "wilderness" where permanent settlement is forbidden, it wasn't always that way.

"I mean, there were people all over this place," said Guenther. We had been overlooking a shallow valley below a pass just shy of eleven thousand feet. The valley had once been filled with wickiups, campfires, and people processing bison from a recent hunt. We had just walked a cairn line used as part of a bison-gifting site (also known as a buffalo jump). The gifting site was used for perhaps eight hundred years, at least up until about two hundred years ago, when colonists actively removed Indigenous people from their homelands.

These are part of the ancestral lands of the Eastern Shoshone. In 1868, the Shoshone were forced onto the Wind River Indian Reservation, situated in the foothills of the eponymous mountain range, along with the Northern Arapaho a decade later. But historically, many tribes traveled through these mountains. Indigenous people subsisted seasonally in the mountains (perhaps some year-round) on an abundance of bison, deer, and bighorn sheep.

Five days into the trip, we'd taken a short side trip up a long windswept ridge to a unique archaeological site revealed to Guenther and his students in 2017. One of Todd's students, Sarah, grew up on the Wind

River Reservation and pointed out an unusual cairn. A cairn is typically built as a stack of rocks with little or no spaces between, but this cairn had an unusual configuration. It consisted of a slab of bedrock on top of which sat two flat stone wedges that leaned against one another, forming a triangular window. "Now," said Guenther, "get down and look through the window and tell me what you see." I laid down on my stomach and peered through the opening. In the middle of the view, I saw the piercing white ice of one of the largest glaciers in the Winds. It filled nearly the entire frame.

"So," Guenther continued, "we thought that was a neat coincidence, but then we kept noodling around (like you do at a site) when we found this one." This cairn was a bit taller and built like a windbreak. On top of the rock wall were three rocks that formed a squarish window. Again, I bent down to peek through the opening, and once again there was the glacier, framed by gray stones covered in lichen.

Although Guenther and his students had seen similar cairns along the established Forest Service trail we'd followed earlier in the trip, they assumed that they were modern features. But when they found these cairns unassociated with the modern trail system and focused on the glacier, they began to wonder if they might be part of a much older and unique tradition.

Over the next several years, Guenther and his students mapped thirty of these sighting features over about twenty-five miles. One sighting cairn directs the viewer's gaze at a perennial ice patch, although Guenther hasn't had the chance to survey it for artifacts. Some of the sighting cairns focused the viewers' attention on other cairns in the distance, which was probably more practical than spiritual. A couple of them contained a tall wooden pole in the center for sighting it at a distance. Samples of wood from one of the poles dated three thousand years old. In all, there are hundreds of trail-marking cairns along the twenty-five-mile route.

"Now," said Guenther, "turn around." I turned and looked back toward what Guenther thinks is the beginning of the pilgrimage. The wide-open golden grasslands of the alpine tundra echoed with the calls of marmots and pika. The view extends for more than a hundred miles. I noticed a swale about thirty feet wide. The swale seemed to follow the line of cairns, like an overgrown footpath. Guenther believes this was a pilgrimage that First Nations people took in celebration of the glaciers. It's not surprising that the people who walked this path and built these sighting cairns understood the value of glaciers as a source of life.

It's hard to say just when or for how long this route was used. Guenther interprets the evidence to suggest that the route was used for several thousand years, sometime between the end of the Pleistocene (around eleven thousand years ago) and the beginning of the Late Precontact Period (about two thousand years ago). I spoke with Crystal Reynolds, an archaeologist for the Northern Arapaho Tribal Historic Preservation Office. She said that in her culture, sighting cairns are common. "You find them all over the place," she told me. But the suspected pilgrimage does not feature in any traditional practices of the Eastern Shoshone or Northern Arapaho, who live on the Wind River Reservation today. Nor does the ethnographic literature of either tribe mention pilgrimages up mountains or the worship of glaciers, although both tribes regard this glacierized basin as sacred. This apparent absence might be because the area was used as a sacred pathway by different groups of Indigenous peoples for many thousands of years before either of these tribes arrived in the area. Or perhaps oral traditions related to this glacier pilgrimage were intentionally erased by colonists who were determined to make the western hemisphere more like western Europe. Sadly, many knowledge holders died of disease, warfare, and genocide during the early decades of the colonial era.

"With oral histories, many of them were not supposed to be written down," Reynolds told me. "Some stories are only supposed to be told at certain times of the year with a particular set of protocols. And those protocols have to be followed in order for that story to be told and shared. If we write it down, then it's general knowledge all the time for anyone." Reynolds said that figuring out a way to capture oral histories is something she and her community are working on now.

* * *

The spirituality of glaciers is felt by cultures all around the world. The glacier-capped summit of Kilimanjaro—the tallest peak in Africa—is of spiritual importance to the local Chagga people. And every year, tens of thousands of Tibetan Buddhists visit two temples located a the foot of the Mingyong Glacier, on the peak Khawa Karpo in the Meili Snow Mountains of Yunnan, China. Anthropologist Elizabeth A. Allison writes that "in venerating mountain deities, local people recognize that their existence is bound up in the prosperity delivered by rivers, which are fed by mountain glaciers and springs." In the Himalaya and Karakoram mountains, locals are finding their own ways to adapt through a highly effective modern take on the ancient practice of procuring glaciers right where they need them.

TEN
___

Glacier Growers and Ancient Knowledge of the Himalayas

Legend has it that a towering wall of ice (like the one that repelled the White Walkers in the book and streaming series *Game of Thrones*) prevented Genghis Khan from invading the Himalaya and Karakoram Mountains in the thirteenth century. The wall of ice is said to have been grown by locals across a narrow mountain pass in an effort to block the relentless advance of the Mongolian empire's army. No one knows how accurate this legend is, but it is rooted in fact. For centuries, people living in the Gilgit-Baltistan region of northern Pakistan have cultivated glaciers by marrying "female" ice to "male" ice. According to locals, the union, if carefully tended, matures into a fully functioning body of ice capable of delivering meltwater to cold desert valleys below.

Gilgit-Baltistan is situated at the confluence of the Hindu Kush, Karakoram, Himalaya, and Pamir mountains. It is home to five of the infamous fourteeners, or peaks over twenty-six thousand feet, including K2, which is the second highest summit in the world. Mountaineers come from all over the world to pit themselves against these giants. Some don't make it home.

Until completion of the Karakoram Highway in 1978, Gilgit-Baltistan was accessible only by foot. The highway undulates over and around mountains along the Indus River from Hasan Abdal in the Punjab province of Pakistan to the Khunjerab Pass in Gilgit-Baltistan, where it crosses into China at 15,439 feet. It is said to be the eighth wonder of the world. Although the location is remote, locals never saw themselves as cut off from the rest of the world, even given the relentless pitch of the extreme topography.

As testament to the connection that mountain passes provide, the Karakoram Highway follows part of the Old Silk Road, on which Chinese silk, tea, spices, jade, and porcelain traveled west and horses, textiles, and glassware traveled east. Traders journeyed parts of the 4000-mile-long Old Silk Road for the 1500 years of the Chinese Han dynasty until 1453, when the Ottoman Empire stopped trading with China. But ancient petroglyphs place people in Gilgit-Baltistan at least four thousand years ago, long before people traveled the Old Silk Road. There are now about two million residents (roughly six hundred people per square mile), surrounded by mountains with more glaciers and ice than anywhere else in the world—yet much of Gilgit-Baltistan is desert-dry.

Aside from the thin green belts tracking the path of glacial meltwater and snowmelt, the adjacent slopes are barren. While snow is often plentiful in the high mountains, intermountain valleys receive minuscule amounts of precipitation. With so little direct rainfall, villagers must capture snowmelt and glacier meltwater for agriculture and day-to-day living. In villages without direct access to glacial meltwater, farmers rely on early-season snowmelt. But spring snowmelt is unpredictable. The amount of meltwater flowing downstream depends on how much snow accumulated the previous winter and how quickly it melts come spring. If the snow melts too quickly, not only might it cause dangerous floods, but much of the water will rush downstream without ever being used. Those living in glacierized basins benefit from not only spring snowmelt but also from late-season glacier melt, but those without glaciers in their watersheds have found ways to adapt.

In the 1940s or 1950s, villagers in Hanuchal (along the Indus River, in west-central Gilgit-Baltistan) grew their own glacier. The climate in Hanuchal at 4700 feet is relatively mild, allowing for two harvest seasons: a late winter wheat crop and a spring corn crop. The growing season ends in mid-autumn when the last of the corn is harvested. Farmers also maintain orchards and raise goats, sheep, and cattle. A system of open-air canals diverts glacial meltwater from a side valley as well as from the glacier grown directly above the village at 15,400 feet.

According to tradition, glaciers are best grown at an altitude of more than thirteen thousand feet, where temperatures average below freezing throughout the year. An ideal spot is out of the direct path of the sun, preferably one that is shadowed for much of the year, especially in summer. Once the location is chosen, villagers dig out a cave from the talus, and inside they place the ingredients needed to grow the glacier. The main ingredient is glacial ice from two specific types of glaciers—female and male.

Twelve men go in search of a female glacier and another twelve men go in search of a male glacier to collect basketfuls of ice. A female glacier is one that is classically glacier-like—white and blue ice with little debris cover. Female glaciers are said to flow more rapidly and give off more water. Male glaciers are debris-covered rock glaciers said to move slowly and yield little water. Rock glaciers are one of the most unique types of glaciers; they are invisible. Hidden below talus, rock glaciers are well insulated from sun, so they melt much more slowly than a typical glacier. They are also resistant to climate change, at least to a certain point. Members of the collecting party may travel long distances to find the right type of ice, even when there is more accessible ice nearby.

The ice is collected in woven willow baskets. Each person carries about sixty pounds of ice down the mountain and then back up to the chosen site. The journey is difficult, making the trek its own kind of pilgrimage. Tradition dictates that no one speaks while carrying the ice, nor should the basket ever touch the ground. If it snows during transport of the ice to the chosen site, it is a good sign for the success of the glacier.

In addition to male and female ice, others collect gourds of water from a local river, which is said to be a mixture of both male and female energy since both types of glaciers contribute to its flow. One glacier grower told anthropologist Ingvar Nørstegård Tveiten, who for his master's thesis at the Norwegian University of Life Sciences documented the practice of growing glaciers, that "water from the river is the most important part of glacier growing. We collect water in the autumn when the river is full of meltwater from the glaciers. Then female and male glacier water is mixed in the river."

Once the ice and gourds of water are set in place, the ice is insulated with some combination of charcoal, sawdust, cloth, wheat husks, and nutshells. Some communities add salt and an aromatic powder called Kāfūr to prevent impurities from entering the ice. After the cave is sealed, the silence is broken with verses from the Quran.

One of the villagers symbolically sacrifices his life for the success of the new glacier. Perhaps as a stand-in for the life of the villager, they sacrifice a goat, distributing its meat to those in need because the act of giving is believed to be important to the success of the new glacier. The ceremony is followed by song and prayer in the village, but the glacier growers don't take a direct path home for fear that the ice will follow them and destroy their village. For the next three years, the glacier is left alone while it takes root. No one is allowed to visit the site and peek at the nascent glacier to check its progress. This incubation period is considered crucial to the infant glacier's success.

Glaciers are often cultivated in avalanche zones, which helps build ice, as snow and rock are piled atop already existing permafrost. Villagers in Gilgit-Baltistan view permafrost as barren or infertile ice. During midsummer, as the seasonal snow melts, water trickles down to the permafrost, where it freezes. Rocks and boulders atop the permafrost help insulate the ice beneath, as with rock glaciers. As the cycle is repeated year after year, the amount of ice grows. The contribution of rock from avalanches can result in the formation of a rock (or male) glacier. A glacier grower in Hanuchal told Tveiten:

> We found a place where water was coming out. This was a sign that there was ice under the soil. We dug down seven feet and found a layer of ice there. It was male ice. We chose to grow the glacier there. Now it has become big and is breaking up the rocks and moving them.

Villagers say that their glacier advanced a half mile down-valley since it was planted. Yet even with all this ice, water is still scarce. Each cluster of households in Hanuchal receives water from the irrigation canal only every eighteen days. Yet, almost all farmers there reported being able to cultivate more land in the half-century since the glacier was planted.

The Ice Men of Ladakh

East of Gilgit-Baltistan and across the border on the Indian side of the Himalayas, civil engineer Chewang Norphel modified the centuries-old technique of glacier growing with the hope of alleviating water shortages in his home of Ladakh. Lying in the rain shadow of the Himalayan watershed, Ladakh is even drier than Hanuchal. There is no summer monsoon season. Fewer than five inches of precipitation fall per year. Temperatures drop to -49°F (-45°C) in winters and hover between 50°F and 86°F (10°C and 30°C) during summer. Despite the harsh environment, people have lived in this vertical desert for thousands of years.

Norphel's idea came from an observation of water leaking from a pipe near his home. As water ran out of the pipe, it flowed at the center. But at the sides, the water moved more slowly and eventually froze into a sheet of ice. Norphel thought that if he could create something similar on a much larger scale, then perhaps he could also grow glaciers.

Norphel constructed his first artificial glacier at Phuktse village in 1987. Since then, he has built fifteen artificial glaciers, earning him the nickname "Ice Man." In Norphel's system, a series of terraces built across a slope diverts water from the main river channel. The diversion channels are then lined with rocks buttressed with local materials—such as manure and soft soils mixed with leaves, sticks, and shrubs—to help strengthen the walls. But the walls are also built to be a bit leaky, so that when the natural glacier above begins to melt, its water is diverted into the channels and seeps through the series of leaky terraces. When temperatures drop, the meltwater forms an entirely new glacier lower down the mountain than the "wild" glacier. Because the artificial glacier is lower than the wild glacier, it melts first, just in time for sowing spring seeds. And just as that water runs out, the "wild" glacier begins to melt, providing a late-season source of water for irrigating crops.

While effective, these artificial glaciers cost money to build and require constant maintenance. A single artificial glacier costs between three thousand and twelve thousand US dollars. Of the ten artificial glaciers created, only two are still being maintained. But innovation often occurs stepwise.

In a twist on this method, Norphel's successor Sonam Wangchuk engineered a new kind of artificial glacier in 2014; he called them "ice stupas." These pyramids of ice are called stupas for their resemblance to traditional Buddhist temples. Ice stupas are created in winter by spraying glacial meltwater straight up into the frigid air. Stupas can be made in winter because, although meltwater streams don't flow as much during winter, the glaciers in the Himalaya and Karakoram mountains are always melting at least a little because of the intense solar radiation of northern India. As the water rains back down, some of it freezes onto a simple structure made of local materials, such as sea buckthorn branches. The technique stores water that would otherwise go unused, since 90 percent of the water used is for agriculture, and farms are not active during winter. Like the artificial glaciers, stupas provide early-season water. Their conical shape exposes stupas to less sunlight than what shines on a flat artificial glacier.

An underground diversion pipe drops two hundred feet in elevation over a mile. Gravitational pressure forces water up through the structure and out of the pipe to rain down and freeze on the branches. The fountain is manually activated at night when temperatures are coldest. As ice forms around the woody shrubs, additional pipes can be added to extend the fountain's height, and more woody plants are added to the ice, which creates an internal structure for the stupa. Stupas can grow up to about two hundred feet tall.

The major advantage of ice stupas over artificial glaciers is that stupas don't need to be located at high altitudes. They also take up very little space, require minimal maintenance, and are comparatively inexpensive to build. Most of the cost of creating an ice stupa comes from laying pipes two to three feet underground so that water does not freeze before it reaches the ice stupa.

These methods merge traditional knowledge with science, with an estimated three- to four-fold increase in income for the farmers they support. There are now more than fifty ice stupas in villages across Ladakh and, like snowflakes, each one is unique. Some are rounded at the top while others are needle-thin. Some have smooth walls while others are draped in stalactites. From the top down, when looking at the stupa itself and the round edge of the fountain's reach, they look like sand dollars. Some of them have grown so large that they last the entire summer and into the following winter.

A sculptor carved a replica of a Buddhist temple out of ice and placed it inside one stupa. It has since become a gathering place for the devout; people travel many miles to pray there. Another stupa has become the site of an ice-climbing competition. And in the Ladakhi district of Leh, villagers have opened a café in their stupa. Stupas have become more than sources of water; they are cultural and social spaces.

Refining Ice Stupas

Suryanarayanan Balasubramanian learned about ice stupas when he volunteered one summer at the Students' Educational and Cultural Movement of Ladakh under the tutelage of Sonam Wangchuk. It was 2015, and he had just earned his master's degree in mathematics. He enjoyed his work there so much that he stayed for three years, serving as the manager for the ice stupa program until heading to Switzerland to pursue his PhD at the University of Fribourg in 2018.

Balasubramanian wanted to understand the factors that make ice stupas grow and how local climate influences ice stupa growth. By the time he began his PhD, Switzerland had its own ice stupa. It was built during the winter of 2016–2017 with the help of Sonam Wangchuk. The following winter, several small ice stupas were built at a site in the Val Morteratsch. But there was a problem common to both ice stupas in Switzerland and those in Ladakh: The pipes sometimes froze. Freezing pipes are a big problem.

When the pipes freeze, the only option is to hope that the pipe isn't damaged, clear it of ice, and start again, repeating the process the next time temperatures fall so low that the pipes freeze again. Balasubramanian hoped to find a solution by studying how ice stupas grow and retain ice in the drastically different climates of Ladakh and Switzerland. What he found surprised him. The Ladakhi stupas grew four times larger than the Swiss stupas. While the Swiss stupas retained about ten thousand liters of water, the Ladakhi stupa retained a million liters. "At first we thought it was a mistake, but we quantified the ice repeatedly over seven or eight ice stupas and always got the same results," said Balasubramanian.

Switzerland is the most mountainous country in Europe, with glaciers spilling down nearly every peak. It is home to some of the best skiing in the world. Why wouldn't Swiss ice stupas grow as large as those in Ladakh? Balasubramanian told me that in Switzerland, a lot of the water sprayed through the fountain wound up pooling on the ground without freezing onto the structure. In Ladakh, though, very little water pooled around stupas. But why? Switzerland is a cold, high-elevation environment with snowy winters. It seemed that it would be one of the best places to grow an ice stupa.

But Balasubramanian told me that when water freezes, a small amount of heat is lost to the atmosphere in a process called sublimation, which occurs when a solid (in this case ice) turns into a gas (water vapor) without passing through its liquid stage. The speed at which this heat is lost determines how much ice is made. Sublimation happens much faster in dry climates and at high altitudes where air pressure is lower than it does in comparatively humid, lower-elevation environments, like Switzerland. The freezing rate for Ladakh ice stupas was more than ten times that of Switzerland's ice stupas. The faster freezing process meant that ice stupas in Ladakh accumulated many times the volume of ice compared to Switzerland's stupas.

Solar radiation also plays a role in ice stupa formation. In winter, solar radiation in Ladakh is twice that in Switzerland, owing to its higher altitude and lower latitude. And while no one would call Switzerland a low country, it is quite a bit lower than Ladakh. In Ladakh, elevations range from 8370 feet to more than 25,000 feet, while in Switzerland elevations range from 633 feet to 15,203 feet. The impact of solar radiation on the Ladakhi ice stupa was half of that in Switzerland. The reason for this has to do with the exposure of the conical structure to both direct and diffuse (scattered by clouds and water vapor) solar radiation. Although less than 20 percent of the structure was exposed to direct solar radiation in both locations, because Switzerland is much cloudier in winter, the amount of diffuse solar radiation is much higher there than in Ladakh.

But even in the favorable climate of Ladakh, where the dry, cold, and cloud-free climate provides the best conditions for growing ice stupas, they are not without problems. Balasubramanian estimates that 75 percent of the water sprayed on ice stupas is lost to evaporation, melting, and water running off the structure rather than freezing. So, his next task was to try and figure out how to make stupas more efficient. One way to do that is to automate the fountain so that it only sprays during the best conditions for making ice.

The automated system that Balasubramanian used required eight times less water than manually turning the fountain on every night. "What this means is that we can build eight structures and get eight times the water supply that a manual ice stupa would produce," said Balasubramanian. An automated fountain system also solves the problem of freezing pipes, since the system senses when it's too cold and shuts down the flow of water. When conditions improve, the system turns the spigot back on.

Although most villagers in Hanuchal, Pakistan, claim that their artificial glacier has increased the amount of arable land, no one has quantified that benefit. Balasubramanian says that putting numbers to artificial ice reservoirs will help communities make the most efficient use of their water. "So, the big question is 'How do we quantify how much water ends up on a farm, and how can farmers use that information to plan irrigation?'"

He is also looking to scale up. Right now, each ice stupa may provide water for a few family farms, but Balasubramanian wants to figure out how to make them large enough to supply an entire village with all the water they need, year in and year out. To accomplish this goal, he founded Acres of Ice. Their mission is to help mountain communities build stupas and find other bespoke solutions to help mitigate water shortages and make the most efficient use of the water they have.

Acres of Ice has inspired another startup based in Santiago, Chile, called Nilus. While ice stupas probably won't help with water shortages in the Alps, the arid, cold, and sunny environment of the Andes is similar enough to the Karakoram and Himalaya mountains that engineers there have begun building stupas and showing others how to build them in their own communities. In the winter of 2021–2022, Nilus founders built Chile's first ice stupa in Parque Arenas in San José de Maipo. This test case proved it is worth trying out elsewhere in the Andes.

Preserving Glaciers

While mountain communities all around the world face many of the same challenges when it comes to melting glaciers, the solutions to water scarcity in mountain environments won't all be artificial glaciers and ice stupas. "We want to use an ensemble of solutions to help them adapt to accelerated decline of glacial meltwater," Balasubramanian told me. Some of the solutions will involve changes to how water is used once it's melted, for instance, employing drip irrigation systems. In the Andes, pre-Spanish irrigation canals, locally known as *amunas*, are being restored by some communities to help solve current water shortages. Some of these canal systems are several thousand years old. They were built by the Wari people, who predate the Inca by many centuries. Amunas are a series of stone canals built across a slope. The canals' serpentine layout slows down the flow of water, which helps it seep into the soil, where it feeds underground springs. These springs serve as critical water sources during the dry season. By storing water underground, amunas avoid water loss due to evaporation, which is high in aboveground water-storage systems like reservoirs.

In Switzerland, scientists have been experimenting with covering parts of ski slopes in white fleece to help preserve the snow and ice. In Switzerland's Upper Engadine, the ablation area (melt zone) of the Diavolezza Glacier, which is part of a ski slope, is covered with polypropylene—a thermoplastic polymer that can be made from 100 percent recycled material and lasts up to five years. Preserving the glacier is important because its presence actually prevents the upper part of a ski run from becoming too steep. And it's working. The covering has led to the thickening of the glacier by about thirty feet over the last decade.

While effective for small glaciers, covering large glaciers with poly-propylene fleece is impractical.

But Felix Keller and Johannes Oerlemans, both glaciologists and musicians, are fighting glacial melt with science and music. Oerle-mans plays double bass guitar and Keller plays violin in their duo TangoGlaciar. Together they've performed in Switzerland and abroad, playing music in honor of melting glaciers. Through music (Keller is also a member of the Swiss Ice Fiddlers), they hope to inspire an emotional response to the loss of mountain ice. And through their nonprofit organization MortAlive, Keller and Oerlemans are experi-menting with ways to preserve glaciers. Their test case is Switzerland's Morteratsch Glacier.

By covering the parts of the glacier most prone to summer melt with a layer of artificial snow, Keller and Oerlemans hope to not only slow down the glacier's retreat but to also encourage new growth. Through model simulations and climate data, Keller and his fellow research-ers found that the addition of just a few inches of snow applied to the glacier's ablation area would be enough to halt the effects of melting within ten to fifteen years. It might even be possible for the glacier to grow. The system will use the glacier's own meltwater to make snow. In Keller's design, free-hanging snow cables strung across the glacier are fed meltwater by gravity from glacial lakes farther up the mountain. The system is called NESSy (New Energy-efficient Snowgun System), and it produces snow without electricity. Keller estimates that if just 10 percent of the glacier is covered in snow, the glacier will stabilize.

While these are intriguing possibilities for mitigating the effects of rapidly melting glaciers, they are also short-term solutions for a long-term problem. Covering vulnerable glaciers in fleece and mak-ing snow to shore them up only buys time. Just how much time is hard to say. At a certain point, it won't matter how much ice is covered in fleece or how much snow is made or how many communities have built artificial ice reservoirs. If average temperatures are too high to sustain perennial ice, then ice will disappear no matter what we do. And once glaciers and perennial ice patches melt, getting them back could take hundreds if not thousands of years. It takes about one hundred thousand years for an ice age to get going and only twenty thousand years to come out of an ice age. While we don't need to head back into a glacial cycle, it's much more effective to maintain the ice we have now than to try and build it back up.

Glacial Lake Outburst Floods

Melting glaciers are not only a problem in terms of water supply; they also pose a dangerous risk to people in some places. As glaciers melt, they often create lakes at their forefields. The lakes are held back by the unconsolidated rock and debris of a glacial moraine, but as the lake continues to fill with meltwater, pressure on the moraine eventually becomes too great. Millions of gallons of water bursts through the natural dam, with catastrophic consequences for those in its path. These time bombs are called glacial lake outburst floods (GLOFs). But Balasubramanian thinks there is a way to defuse these bombs while also storing water for locals.

In 2022, one such flood occurred in Gilgit-Baltistan. A May heat wave swept across the region, causing rapid and extensive melt of the Shisper Glacier. When a lake at the glacier's base burst through its moraine, a massive flood destroyed the Hassanabad bridge on the Karakoram Highway. Although GLOFs occur every year (averaging five or six per year), in 2022 there were at least sixteen. As if the Shisper GLOF wasn't enough, the May heat wave coupled with extreme monsoon rains a month later led to monster flooding that affected the entire nation in August 2022. The flood displaced 32 million people and caused 1700 deaths, and one third of the victims were children.

Since 1990, the number of glacial lakes has risen. At the same time, the populations living in their paths have also increased. One study estimates that at least fifteen million people worldwide are at risk of experiencing a GLOF. While GLOFs are most common in Asia's mountains (Pakistan has more than three thousand glacial lakes), they also occur in the Andes, Canadian Rockies, and Alps.

The Politics of Ice

The Siachen Glacier is the second longest in the world, originating at 18,900 feet and terminating forty-seven miles later at 11,900 feet. It originates on a mountain pass on the border of India and China. The glacier's melting waters are the main source of the Nubra River, which is located in the Indian region of Ladakh and drains into the Shyok River. The Shyok in turn joins the 1800-mile-long Indus River, which flows through Pakistan. Thus, the glacier is a major source of the Indus and feeds the largest irrigation system in the world. It is also the site of the world's highest battleground.

The World Economic Forum says that future wars will be fought over water. The five most vulnerable hot spots are the Nile river, the Ganges-Brahmaputra Delta, the Indus river, the Tigris-Euphrates river system, and the Colorado river. Although both Pakistan and India cite water as one reason to fight over the Siachen Glacier, as Indian Lieutenant-General V. R. Raghavan wrote, the "Siachen has become embedded in the Indian public consciousness as a symbol of national will" of post-colonial reclamation.

Gilgit-Baltistan and Ladakh are part of the contested Kashmir region, which also includes the India-administered territories of Jammu, Kashmir, and Ladakh; the Pakistan-administered territories of Azad Jammu and Kashmir; and the China-administered territories of Aksai Chin and the Trans-Karakoram Tract. India and Pakistan have fought over Kashmir since 1947, when British crown rule of the region abruptly ended. When Pakistan and India signed the Karachi Agreement in

1949, the United Nations failed to precisely demarcate the border line across the Siachen Glacier. The Line of Control established in 1972 ran through Kashmir to a spot in the Karakoram, an indistinguishable coordinate still miles south of the Chinese border and known simply as NJ9842. The agreement described the border as running "north to the glaciers"—a nebulous description that would lead to the outbreak of conflict.

United Nations officials never imagined that a place so rugged and high in altitude would be contested, but the region became a flashpoint of national identity and pride. Neither Pakistan nor India is willing to cede this small, inhospitable, and remote landscape. Both countries support troops at outposts situated atop the glacier at more than twenty-one thousand feet—fewer than five thousand feet below the death zone. Far more people have died of slips into crevasses, cerebral edema, ice falls, and avalanches than have died in battle. In 2012, an avalanche killed 129 Pakistani soldiers and eleven civilians. An anonymous Kashmiri poet wrote the following piece about the conflict.

> *I cannot drink water*
>
> *It is mingled with the blood of young men who have died up in the mountains.*
>
> *I cannot look at the sky*
>
> *It is no longer blue; but painted red.*
>
> *I cannot listen to the roar of the gushing stream*
>
> *It reminds me of a wailing mother next to the bullet-ridden body of her only son.*
>
> *I cannot listen to the thunder of the clouds*
>
> *It reminds me of a bomb blast.*
>
> *I feel the green of my garden has faded*

Perhaps it too mourns.

I feel the sparrow and cuckoo are silent

Perhaps they too are sad.

—Anonymous Kashmiri poet

Glaciers are perceived as pristine environments not yet polluted by pesticides, human waste, animal waste, and other contaminants. Yet, glaciers provide some of the best information on both local (mountain glaciers) and global (polar ice sheets) sources of pollution. Some of the most common pollutants that accumulate in ice are black carbon, microplastics, fallout radionuclides, nitrates from agriculture and farming, pesticides, and pharmaceuticals. When glaciers and ice sheets melt, they release these contaminants downstream and into the oceans. Remember that meltwater from mountain glaciers provides one third of the world's population with drinking water, and the oceans are a source of food for people worldwide. After decades of military occupation atop the Siachen Glacier, the upper Indus has tested positive for heavy metals, antibiotics, human waste, and toxic residue from the camps.

* * *

While Himalayan glaciers are receding, there is still a great deal of ice in these mountains. So much ice, in fact, that it's hard to imagine the mountains without their white shrouds. In the Sherpa scriptures, there is a prophecy warning that inattention to the Buddha Dharma will bring about the apocalypse, in which nine suns will scorch Earth. It's a prophecy that eerily parallels the current climate-change crisis and shares similarities with the Quechua prophecy of darkening mountain peaks when the apus are not given proper respect.

In the Himalayas, it's believed that deities inhabit mountaintops (as it is in the Andes). Tibetan Buddhists in Ladakh prohibit cooking or eating foods such as garlic and onions; burning meat; and experiencing strong emotions, breaking vows, or physically fighting at high elevations near glaciers; similar prohibitions exist in Tlingit culture. Preventing pollution, either spiritual or physical, protects the mountains. When they are disrespected, bad things happen. Glacial lake outburst floods. Drought. Unpredictable river flows. Extinction of ice. Disrespect and imbalance are behind climate change, whether we attribute it to inattention to the gods or impersonal human negligence. Like a serpent swallowing its own tail, we have trapped ourselves in an ouroboros of self-destruction. And yet, our fate is not sealed. We can harness our ingenuity and adaptability to engage in meaningful change.

ELEVEN

The Future of Ice-Patch Archaeology

Along the Siberian border, nestled alongside the large Darkhad Depression, lies a vast landscape of tundra and larch forest. The Ulaan Taiga Specially Protected Areas are home to the reindeer-herding Dukha people (ethnic Tuvans). The Dukha are the most southern-latitude reindeer-herders in the world and may have been the first people to domesticate the species. Today, the Dukha comprise a group of around two hundred people, or about thirty families. Most families herd twenty to thirty reindeer, but some families may herd as many as a hundred animals. They raise reindeer for their rich, fatty milk, which they use to make butter, yogurt, milk tea, and cheese.

Archaeologist William Taylor, assistant professor and curator of archaeology at the University of Colorado in Boulder, studies the enduring relationship between humans and domestic animals in Mongolia and elsewhere. Reindeer's reliance on ice patches, coupled with the emergence of the field of ice-patch archaeology, inspired Taylor to explore the archaeological potential of ice patches in northern Mongolia.

Through conversations with Dukha families in 2016 and 2017, Taylor learned that several of the ice patches that the reindeer use to cool off and avoid parasites had melted for the first time in the families' collective memory. In 2018, Taylor searched for artifacts at eleven ice patches that hadn't yet melted. At one ice patch, he found a stick similar to the scaring sticks found in Norway. He also found two pieces of a willow branch with beveled ends; this design is one traditionally used by the Dukha because the method creates a strong yet flexible joint. The local people thought it might be part of a fishing pole. Although these pieces are not that old, dating to the 1960s or 1970s, they point to the recent loss of ice in a place where ice had been permanent for at least half a century, but probably much longer. When Taylor returned the following year, all eleven ice patches had disappeared. Heat-stressed reindeer now lie in the dirt. The traditional lifeway of the Dukha reindeer-herders is imperiled.

The loss of ice is more than physical; it has implications for how cultures see themselves. The Dukha and Sami reindeer-herders are acutely aware of how the loss of ice affects their livelihoods, their cultural traditions, and the health of their animals. Similarly, those living in the European Alps are also facing loss of livelihoods and cultural identity as their glaciers melt. Quechua farmers wonder what will happen to the apus when the glaciers are gone. What will Icelanders call themselves when their glaciers melt into the ocean?

Ice-patch archaeology can't stop climate change, but it can illuminate the cultural implications of melting ice and aid in the preservation of cultural material that would otherwise be lost. Concerned about the loss of ice in the traditional lands of the Dukha reindeer-herders, Taylor visited the Altai Mountains of western and northwestern Mongolia to document the archaeological potential of ice patches there before they, too, disappear.

About five thousand years ago, migrants from eastern Europe brought sheep, goats, and cattle to the Altai. These people lived a semi-nomadic lifestyle similar to that of the Dukha and Sami reindeer-herders. By the late Bronze Age, horses had transformed the economy and culture of northwestern Mongolia, allowing for greater accessibility into the mountains. Once horses were introduced, they became an integral part of nomadic cultures in the region. Most archaeologists had assumed that, given the long history of pastoralism and the large volume of domestic animal bones found at more traditional archaeological sites, herders did not hunt wild game. But the discoveries Taylor made in 2019 challenge this assumption.

Taylor surveyed five ice patches and a glacier spilling down the east face of Tsengel Khairkhan, which rises nearly thirteen thousand feet on the east side of the Altai. As we've learned, glaciers aren't typically included in ice-patch surveys, but locals told Taylor that climbers had found wood and antler projectiles at this site in previous years, so he added it to his list of sites to check.

By the end of August 2019, Taylor had collected artifacts that span the last 3500 years. In a trench between the glacier's ice margin and a rock wall, he found almost two dozen wooden arrow shafts and a couple of dart fragments made of willow. One arrow shaft was intact, tipped with a bronze point, and hafted with animal sinew. He also found several iron and bone arrowheads made of red deer antler and argali sheep bones. Bronze Age pastoralists chose to make some of their hunting tools with wild animal bone, antler, and sinew even though it would have been easier to use bone and sinew from their domestic animals. Perhaps because they viewed wild sheep and deer bones as finer material.

Taylor also found several sticks that had been sharpened at one end. He thinks that the glacier, and that spot in particular, could have been used to drive wild sheep onto the ice in the same way that Norwegian hunters used scaring sticks. Taylor found dozens of horn sheaths and horn cores along the northern edge of the glacier near the mountain's summit, confirming that that area was used to hunt wild sheep. As Taylor continued his surveys, he noticed dozens more sheaths and cores poking out of the glacier. He told me that hundreds of wild argali sheep were probably harvested at this site, but only their heads remained.

Hunters may have removed the heads in part to lighten the carcasses for transport back down to camp. But beyond the practicality of leaving behind heavy sheep skulls, they appeared to be intentionally placed, although Taylor admits that it's possible the skulls were moved in more recent times. In Mongolia, the head of an animal is imbued with special significance. For example, the skulls of deceased horses are often placed at high-mountaintop stone cairns called *ovoos*. During the late Bronze Age, the ritual placement of the head, neck, and hoof bones of deceased horses was standard practice at burial and monument sites. Hunting wild argali sheep and red deer may have been part of a spiritual practice as well as a means of subsistence. Using their bones, horns, and sinew may have been spiritual as well.

While the glacier was used to hunt wild sheep, Taylor discovered that the lower-elevation ice patches were used to pasture domestic sheep and goats, both in recent years as well as in the distant past (two thousand to three thousand years ago). The ice patches and glacier provide the first and only evidence of wild-game hunting in early pastoral cultures of Mongolia. If Ötzi hadn't melted out of the Ötztal Alps or the hundreds of arrows and atlatls hadn't been found at the melting edges of ice in the Yukon, Taylor might never have thought to look at ice patches in Mongolia. The evidence of this unique hunting tradition could have been lost.

How Can We Be Good Ancestors?

Most of the world's mountains have yet to be surveyed for ice-patch artifacts. New Zealand, Iceland, Sweden, the Caucasus and Middle East, and North Asia are all places where ice patches may yield clues to our past. In traditional archaeology, one can choose whether or not to open a grave or excavate a dwelling. If left alone, an archaeological site buried in the ground is relatively safe, provided the area aboveground isn't slated for development. Ice-patch archaeologists do not have the luxury of time. If the cultural material isn't collected from ice patches, their soft organic parts will disintegrate, and when the ice inevitably melts, all that will be left are the stones and bones of traditional archaeology. Important stories of these objects will never get told to living ancestors. If no one is looking now, we'll never know what was lost.

The task assumed by ice-patch archaeologists is monumental. There are only a handful of such archaeologists worldwide, yet there are tens of thousands of perennial ice patches with archaeological potential in mountain ranges around the world. There just isn't enough time.

The rapidity and scale of warming means that today's ice-patch archaeologists have a huge responsibility to tomorrow's archaeologists. After the ice is gone, archaeologists will be limited to studying archived collections. They will have to rely entirely on the information and material that ice-patch archaeologists are collecting today. There will be no ice patch to return to, yet the questions will remain. What we choose to do today—the actions that governments, corporations, and we as individuals take—matters for tomorrow. We must be as protective of our future as we are curious about our past.

As scientist Jonas Salk, who developed the life-saving polio vaccine in 1955, said: "Our greatest responsibility is to be good ancestors." We've inherited the world of our ancestors—the sum total of what they did or didn't do and how that affects the political, social, economic, and environmental circumstances of the world we live in today. Our actions or inaction will create the world our descendants inhabit. Perhaps more than any other field of scientific study, ice-patch archaeology brings awareness to climate change. We can see the ice melting in real time. We understand that humans are largely responsible, but it's hard to take responsibility. Being a good ancestor means planning for the future even though we won't bear witness to it.

In 2014, Iceland's Okjökull glacier was pronounced dead when its ice was no longer thick enough to move. A plaque installed at the site reads: "In the next 200 years, all our glaciers are expected to follow the same path. This monument is to acknowledge that we know what is happening and what needs to be done. Only you know if we did it."

I think of you, reading this book, maybe after it was found by a hiker at the site of the last perennial ice patch that somehow survived, thousands of years from now. Perhaps these words were reconstructed, or perhaps they were frozen and perfectly preserved, the cover still intact, albeit waterlogged. Perhaps only the gist of the book's message can be retained, but enough to translate the crisis of our world's melting. Enough to ask: Did we accomplish what needed to be done?

Acknowledgments

First and foremost, thank you to the dozens of scientists and cultural resource managers who gave freely of their time to help me understand ice-patch archaeology, glacier mechanics, and cultural ties to ice patches and glaciers. Of particular note, I thank Craig Lee, Joseph McConnell, Nathan Chellman, Diane Strand, Lawrence Joe, Kelsey Pennanen, Christian Thomas, Greg Hare, Lars Pilø, Linda Jarrett, Rune Odegard, and Albert Hafner. I also thank Apus Peru Adventure Travel Specialists for such wonderful support on the Qoyllur Rit'i trek and for putting up with all my follow-up questions in the months afterward. A special thanks to Todd Guenther, Lucas, Hannah Nelson, Cadence Trusho, Allison Stoff, and Rita for never making me feel bad about how slowly I hiked on our fourteen-day trek in the Winds. Lucas, you are an excellent crew leader, medic, and camp companion. I'm so grateful for your company.

I'd also like to thank my wonderful husband, John. Were it not for his support, this book would never have been. Your patience and encouragement helped keep me sane, especially during the messy middle when I thought that maybe I'd never finish writing this book. I also thank my most trusted reader, my mother, Karen Baril, for reading countless drafts and still encouraging me to keep going. You cheered me on like a mother but advised me like the editor and writer you are. You provided exactly the kind of tough love I needed. I also thank my father, David Baril, for his behind-the-scenes support and encouragement.

I also thank the Communicating the Climate Crisis writing group. You all are amazing, talented writers. Spending a week in Chamonix with you all is one of my most treasured memories. Thank you all for your support and encouragement. I'm especially grateful to Lauren E. Oakes and Emily Polk, who led the writing group. Thank you to E. James Dixon, Tonya Opperman, and Hali Kirby, who provided invaluable comments on earlier drafts.

Last but not least, it takes a whole team to put together a book. Writing a book may seem like a solo endeavor, but I can assure you, it is not. I thank my wonderful agent Jessica Papin, who took a chance on an unknown writer with a great idea. Thank you, Jessica! I also thank my editor Makenna Goodman, whose suggestions made this book better; acquiring editor Stacee Gravelle Lawrence; copy editor Ellen Foreman; production editor Mathew Burnett, and the rest of the production team at Timber Press. Any mistakes are inadvertent and my own.

Selected Bibliography

Ackermann, Brandon, Craig M. Lee, David McWethy, Nathan Chellman, and Joe McConnell. 2021. "An Application of Ground-Penetrating Radar at a Greater Yellowstone Area Ice Patch." *Journal of Glacial Archaeology* 5: 73–84.

Allison, Elizabeth A. 2015. "The Spiritual Significance of Glaciers in an Age of Climate Change." *Wiley Interdisciplinary Reviews: Climate Change* 6, no. 5: 493–508.

Andrews, Thomas D., Glen MacKay, Leon Andrew, Wendy Stephenson, Amy Barker, Claire Alix, and the Shúhtagot'ine Elders of Tulita. 2012. "Alpine Ice Patches and Shúhtagot'ine Land Use in the Mackenzie and Selwyn Mountains, Northwest Territories, Canada." *ARCTIC* 65: 22–42.

Balasubramanian, Suryanarayanan. "Sustaining Glacial-Fed Catchments with Artificial Ice Reservoirs." Public thesis presentation at the University of Fribourg for the award of the PhD in Geosciences, Fribourg, Switzerland, May 5, 2023.

Behringer, Wolfgang. 1999. "Climatic Change and Witch-Hunting: The Impact of the Little Ice Age on Mentalities." *Climatic Change* 43: 335–351.

Birkhold, Matthew H. 2019. "Measuring Ice: How Swiss Peasants Discovered the Ice Age." *The Germanic Review: Literature, Culture, Theory* 94: 194–208.

Boers, Bernice de Jong. 1995. "Mount Tambora in 1815: A Volcanic Eruption in Indonesia and its Aftermath. *Indonesia* 60: 37–60.

Britton, Kate and Charlotta Hillerdal. 2019. "Archaeologies of Climate Change: Perceptions and Prospects." *Études Inuit Studies* 43: 265–288.

Burke, Ariane, Matthew C. Peros, Colin D. Wren, Francesco S. R. Pausata, Julien Riel-Salvatore, Olivier Moine, Anne de Vernal, Masa Kageyama, and Solène Boisard. 2021. "The Archaeology of Climate Change: The Case for Cultural Diversity." *Proceedings of the National Academy of Sciences* 118, no. 30: e2108537118.

Callanan, Martin. 2014. "Out of the Ice: Glacial Archaeology in Central Norway." PhD Thesis, Norwegian University of Science and Technology.

Carey, Mark. 2007. "The History of Ice: How Glaciers Became an Endangered Species." *Environmental History* 12, no. 3: 497–527.

Ceruti, Maria Constanza. 2004. "Human Bodies as Objects of Dedication at Inca Mountain Shrines (North-Western Argentina)." *World Archaeology* 36, no. 1: 103–122.

Ceruti, Maria Constanza. 2015. "Frozen Mummies from Andean Mountaintop Shrines: Bioarchaeology and Ethnohistory of Inca Human Sacrifice." *BioMed Research International* 6: 1–12.

Chellman, Nathan J., Gregory T. Pederson, Craig M. Lee, David B. McWethy, Kathryn Puseman, Jeffery R. Stone, Sabrina R. Brown, and Joseph R. McConnell. 2021. "High Elevation Ice Patch Documents Holocene Climate Variability in the Northern Rocky Mountains." *Quaternary Science Advances* 3: 100021.

Cruikshank, Julie. 2005. *Do Glaciers Listen?: Local Knowledge, Colonial Encounters, and Social Imagination*. Vancouver, BC, Canada: University of British Columbia Press.

ACKNOWLEDGMENTS

Dixon, E. James, William F. Manley, and Craig M. Lee. 2005. "The Emerging Archaeology of Glaciers and Ice Patches: Examples from Alaska's Wrangell–St. Elias National Park and Preserve." *American Antiquity* 70, no. 1: 129–143.

Dixon, E. James. 2013. *Arrows and Atl Atls: A Guide to the Archeology of Beringia.* Washington, D.C.: United States Department of the Interior.

Dixon, E. James, Martin Eugene Callanan, Albert Hafner, and P. G. Hare. 2014. "The Emergence of Glacial Archaeology." Journal of Glacial Archaeology 1: 1–9.

Earle, Steven. 2021. A Brief History of the Earth's Climate: Everyone's Guide to the Science of Climate Change. Gabriola Island, BC, Canada: New Society Publishers.

Fagan, Brian. 2000. The Little Ice Age: How Climate Made History 1300–1850. New York, NY: Basic Books.

Farbregd, Oddmunn. 1972. "Arrow Finds from the Mountains of Oppdal." University of Trondheim: The Royal Norwegian Society of Sciences Museum.

Fowler, Brenda. 2001. Iceman: Uncovering the Life and Times of a Prehistoric Man Found in an Alpine Glacier. 2nd ed. Chicago, IL: The University of Chicago Press.

Gagné, Karine, Mattias Borg Rasmussen, and Ben Orlove. 2014. "Glaciers and Society: Attributions, Perceptions, and Valuations." *Wiley Interdisciplinary Reviews: Climate Change* 5, no. 6: 703–848.

Gagné, Karine. 2016. "Cultivating Ice Over Time: On the Idea of Timeless Knowledge and Places in the Himalayas." *Anthropologica* 58, no. 2: 193–210.

Giles, Melanie. 2009. "Iron Age Bog Bodies of North-western Europe: Representing the Dead." *Archaeological Dialogues* 16, no. 1: 75–101.

Glave, Edward J. 2013. "Travels to the Alseck: Edward Glave's Reports from Southwest Yukon and Southeast Alaska, 1890-91." Edited by Julie Cruikshank, Doug Hitch, and John Ritter. Whitehorse, Yukon, Canada: Yukon Native Language Centre.

Gostner, Paul, and Eduard Egarter Vigl. 2002. "INSIGHT: Report of Radiological-Forensic Findings on the Iceman." *Journal of Archaeological Science* 29, no. 3: 323–326.

Gostner, Paul, Patrizia Pernter, Giampietro Bonatti, Angela Graefen, and Albert R. Zink. 2011. "New Radiological Insights into the Life and Death of the Tyrolean Iceman." *Journal of Archaeological Science* 38, no. 12: 3425–3431.

Gräslund, Bo, and Neil Price. 2012. "Twilight of the Gods? The 'Dust Veil Event' of AD 536 in Critical Perspective." *Antiquity* 86, no. 332: 428–443.

Greer, Sheila, and Diane Strand. 2012. "Cultural Landscapes, Past and Present, and the South Yukon Ice Patches." *ARCTIC* 65 (Supplement 1): 136–152.

Grize, Leticia, Anke Huss, Oliver Thommen, Christian Schindler, and Charlotte Braun-Fahrländer. 2005. "Heat Wave 2003 and Mortality in Switzerland." *Swiss Medical Weekly* 135, nos. 13-14: 200-205.

Grosjean, Martin, Peter J. Suter, Mathias Trachsel, and Heinz Wanner. 2007. "Ice-Borne Prehistoric Finds in the Swiss Alps Reflect Holocene Glacier Fluctuations." *Journal of Quaternary Science* 22, no. 3: 203–207.

Grove, Jean M. 2004. *Little Ice Ages: Ancient and Modern.* 2nd ed, Volume I. New York, NY: Routledge.

Hafner, Albert. 2012. "Archaeological Discoveries on Schnidejoch and at Other Ice Sites in the European Alps." *ARCTIC* 65 (Supplement 1): 189–202.

Hafner, Albert. 2015. "Schnidejoch and Lötschenpass: Archaeological Research in the Bernese Alps, Volume 1." Bern, Switzerland: Archaeological Service of the Canton Bern.

Hafner, Albert, and Christoph Schwörer. 2018. "Vertical Mobility around the High-Alpine Schnidejoch Pass: Indications of Neolithic and Bronze Age Pastoralism in the Swiss Alps from Paleoecological and Archaeological Sources." *Quaternary International* 484: 3–18.

Hamblyn, Richard. 2009. "The Whistleblower and the Canary: Rhetorical Constructions of Climate Change." *Journal of Historical Geography* 35, no. 2: 223–236.

Hare, P. Gregory, Sheila Greer, Ruth Gotthardt, Richard Farnell, Vandy Bowyer, Charles Schweger, and Diane Strand. 2004. "Ethnographic and Archaeological Investigations of Alpine Ice Patches in Southwest Yukon, Canada." *ARCTIC* 57, no. 3: 260–272.

Hare, P. Gregory, Christian D. Thomas, Timothy N. Topper, and Ruth M. Gotthardt. 2012. "The Archaeology of Yukon Ice Patches: New Artifacts, Observations, and Insights." *ARCTIC* 65 (Supplement 1): 118–135.

Hebda, Richard J., Sheila Greer, and Alexander P. Mackie. 2017. *Kwäday Dän Ts'ìnchį: Teachings from Long Ago Person Found*. Victoria, BC: Royal British Columbia Museum.

Herbert, Timothy D. 2023. "The Mid-Pleistocene Climate Transition." *Annual Review of Earth and Planetary Sciences* 51: 389–418.

Hock, Regine, Golam Rasul, Carolina Adler, Bolivar Cáceres, Stephan Gruber, Yukiko Hirabayashi, Miriam Jackson, Andreas Kääb, Shichang Kang, Stanislav Kutuzov, Alexander Milner, Ulf Molau, Samuel Morin, Ben Orlove, Heidi Steltzer, et. al. 2019. "High Mountain Areas." In *IPCC Special Report on the Ocean and Cryosphere in a Changing Climate*, edited by Hans-Otto Pörtner, Debra Roberts, Valerie Masson-Delmotte, Panmao Zhai, M. Tignor, Elvira Poloczanska, Katja Mintenbeck, A. Alegría, Maike Nicolai, Andrew Okem, Jan Petzold, B. Rama, and Nora M. Weyer, 131–202. Cambridge, UK and New York, NY, USA: Cambridge University Press.

Hougen, Bjørn. 1937. "The Arrows from Storhø." *Viking* 1: 97–204.

Hughes, Susan S. 1998. "Getting to the Point: Evolutionary Change in Prehistoric Weaponry." *Journal of Archaeological Method and Theory* 5: 345–408.

Hugonnet, Romain, Robert McNabb, Etienne Berthier, Brian Menounos, Christopher Nuth, Luc Girod, Daniel Farinotti, Matthias Huss, Ines Dussaillant, Fanny Brun, and Andreas Kääb. 2021. "Accelerated Global Glacier Mass Loss in the Early Twenty-First Century." *Nature* 592: 726–731.

Ion, Peter G., and G. Peter Kershaw. 1989. "The Selection of Snowpatches as Relief Habitat by Woodland Caribou (*Rangifer tarandus* caribou), Macmillan Pass, Selwyn/Mackenzie Mountains, N.W.T., Canada." *Arctic and Alpine Research* 21, no. 2: 203–211.

IPCC. 2023. Climate Change 2023: Synthesis Report. Contribution of Working Groups I, II and III to the Sixth Assessment Report of the Intergovernmental Panel on *Climate Change* (Core Writing Team, H. Lee and J. Romero [eds.]). IPCC, Geneva, Switzerland: 184.

Jackson, M. 2015. "Glaciers and Climate Change: Narratives of Ruined Futures." Wiley Interdisciplinary Reviews: Climate Change 6, no. 5: 479–492.

Jarrett, Linda. 2018. "Into the Ice: A Study of Glaciological and Geomorphological Characteristics of Archeologically Significant Ice Patches in Central Norway." PhD Thesis, Norwegian University of Science and Technology.

Junkmanns, Jürgen, Johanna Klügl, Giovanna Di Pietro, and Albert Hafner. 2021. "The Neolithic Bow Case from Lenk, Schnidejoch: A Technological and Cultural Analysis." *Journal of Glacial Archaeology* 5: 5–50.

Kania, Marta. 2019. "The Qoyllurit'i Pilgrimage: Religious Heritage versus Socio-Environmental Problems." *Studia Religiologica* 52, no. 3: 205–220.

Krüger, Tobias. 2013. "Discovering the Ice Ages: International Reception and Consequences for a Historical Understanding of Climate." Leiden, South Holland: Brill.

Kuzyk, Gerald W., Donald E. Russell, Richard S. Farnell, Ruth M. Gotthardt, P. Gregory Hare, and Erik Blake. 1999. "In Pursuit of Prehistoric Caribou on Thandlät, Southern Yukon." *ARCTIC* 52, no.2: 214–219.

Ladurie, Emmanuel Le Roy. 1971. *Times of Feast, Times of Famine: A History of Climate Since the Year 1000*. Garden City, NY: Doubleday & Company.

Lee, Craig M. 2012. "Withering Snow and Ice in the Mid-latitudes: A New Archaeological and Paleobiological Record for the Rocky Mountain Region." *ARCTIC* 65 (Supplement 1): 165–177.

Lee, Craig M., and Kathryn Puseman. 2017. "Ice Patch Hunting in the Greater Yellowstone Area, Rocky Mountains, USA: Wood Shafts, Chipped Stone Projectile Points, and Bighorn Sheep (*Ovis canadensis*)." *American Antiquity* 82, no. 2: 223–243.

Macdougall, Doug. 2004. *Frozen Earth: The Once and Future Story of Ice Ages*. 1st ed. Oakland, California: University of California Press.

Maixner, Frank, Dmitrij Turaev, Amaury Cazenave-Gassiot, Marek Janko, Ben Krause-Kyora, Michael R. Hoopmann, Ulrike Kusebauch, Mark Sartain, Gea Guerriero, Niall O'Sullivan, Matthew Teasdale, Giovanna Cipollini, Alice Paladin, Valeria Mattiangeli, Marco Samadelli, Umberto Tecchiati, Andreas Putzer, Mine Palazoglu, John Meissen, Sandra Lösch, Philipp Rausch, John F. Baines, Bum Jin Kim, Hyun-Joo An, Paul Gostner, Eduard Egarter-Vigl, Peter Malfertheiner, Andreas Keller, Robert W. Stark, Markus Wenk, David Bishop, Daniel G. Bradley, Oliver Fiehn, Lars Engstrand, Robert L. Moritz, Philip Doble, Andre Franke, Almut Nebel, Klaus Oeggl, Thomas Rattei, Rudolf Grimm, and Albert Zink. 2018. "The Iceman's Last Meal Consisted of Fat, Wild Meat, and Cereals." *Current Biology* 28, no. 14: 2348–2355.e9.

Miles, Martin W., Camilla S. Andresen, and Christian V. Dylmer. 2020. "Evidence for Extreme Export of Arctic Sea Ice Leading the Abrupt Onset of the Little Ice Age." *Science Advances* 6, no. 38: eaba4320.

Miller, Gifford H., Áslaug Geirsdóttir, Yafang Zhong, Darren J. Larsen, Bette L. Otto-Bliesner, Marika M. Holland, David A. Bailey, Kurt A. Refsnider, Scott J. Lehman, John R. Southon, Chance Anderson, Helgi Björnsson, and Thorvaldur Thordarson. 2012. "Abrupt Onset of the Little Ice Age Triggered by Volcanism and Sustained by Sea-ice/Ocean Feedbacks." *Geophysical Research Letters* 39, no. 2: L02708.

Munir, Ramsha, Tehzeeb Bano, Iftikhar Hussain Adil, and Umer Khayyam. 2021. "Perceptions of Glacier Grafting: An Indigenous Technique of Water Conservation for Food Security in Gilgit-Baltistan, Pakistan." *Sustainability* 13, no. 9: 5208.

Murphy Jr., William A., Dieter zur Nedden, Paul Gostner, Rudolph Knapp, Wolfgang Recheis, and Horst Seidler. 2003. "The Iceman: Discovery and Imaging." *Radiology* 226, no. 3: 614–629.

Nardin, Jane. 1999. "A Meeting on the Mer de Glace: *Frankenstein* and the History of Alpine Mountaineering." *Women's Writing* 6, no. 3: 441–449.

Nesje, Atle, Lars H. Pilø, Espen Finstad, Brit Solli, Vivian Wangen, Rune S. Ødegård, Ketil Isaksen, Eivind N. Støren, Dag Inge Bakke, and Liss M. Andreassen. 2012. "The Climatic Significance of Artefacts Related to Prehistoric Reindeer Hunting Exposed at Melting Ice Patches in Southern Norway." *The Holocene* 22, no. 4: 485–496.

Ødegård, Rune S., Atle Nesje, Ketil Isaksen, Liss Marie Andreassen, Trond Eiken, Margit Schwikowski, and Chiara Uglietti. 2017. "Climate Change Threatens Archaeologically Significant Ice Patches: Insights into Their Age, Internal Structure, Mass Balance and Climate Sensitivity." *The Cryosphere* 11, no. 1: 17–32.

Oeggl, Klaus, Werner Kofler, Alexandra Schmidl, James H. Dickson, Eduard Egarter-Vigl, and Othmar Gaber. 2007. "The Reconstruction of the Last Itinerary of "Ötzi", the Neolithic Iceman, by Pollen Analyses from Sequentially Sampled Gut Extracts." *Quaternary Science Reviews* 26: 853–861.

Oerlemans, Johannes, Suryanarayanan Balasubramanian, Conradin Clavuot, and Felix Keller. 2021. "Brief Communication: Growth and Decay of an Ice Stupa in Alpine Conditions – A Simple Model Driven by Energy-Flux Observations over a Glacier Surface." *The Cryosphere* 15, no. 6: 3007–3012.

Oerlemans, Johannes, Martin Haag, and Felix Keller. 2017. "Slowing Down the Retreat of the Morteratsch Glacier, Switzerland, by Artificially Produced Summer Snow: A Feasibility Study." *Climatic Change* 145, no. 3: 189–203.

O'Sullivan, Niall J., Matthew D. Teasdale, Valeria Mattiangeli, Frank Maixner, Ron Pinhasi, Daniel G. Bradley, and Albert Zink. 2016. "A Whole Mitochondria Analysis of the Tyrolean Iceman's Leather Provides Insights into the Animal Sources of Copper Age Clothing." *Scientific Reports* 6: 31279.

Painter, Thomas H., Mark G. Flanner, Georg Kaser, Ben Marzeion, Richard A. VanCuren, and Waleed Abdalati. 2013. "End of the Little Ice Age in the Alps Forced by Industrial Black Carbon." *Proceedings of the National Academy of Sciences* 110, no. 38: 15216–15221.

Paerregaard, Karsten. 2020. "Searching for the New Human: Glacier Melt, Anthropogenic Change, and Self-reflection in Andean Pilgrimage." *Journal of Ethnographic Theory* 10, no. 3: 844–859.

Pernter, Patrizia, Paul Gostner, Eduard E. Vigl, and Frank J. Rühli. 2007. "Radiologic Proof for the Iceman's Cause of Death (ca. 5300BP)." *Journal of Archaeological Science* 34, no. 11: 1784–1786.

Pilø, Lars, Brit Solli, Elling Utvik Wammer, and Vivian Wangen. 2010. "The Archaeological Survey at the Juvfonna Ice Patch, Lom, Oppland, Norway." Report presented at Oppland County Municipality and Museum of Cultural History.

Pilø, Lars, Espen Finstad, and James H. Barrett. 2020. "Crossing the Ice: An Iron Age to Medieval Mountain Pass at Lendbreen, Norway." *Antiquity* 94: 437–454.

Pilø, Lars, Espen Finstad, Christopher B. Ramsey, Julian R. P. Martinsen, Atle Nesje, Brit Solli, Vivian Wangen, Martin Callanan, and James H. Barrett. 2018. "The Chronology of Reindeer Hunting on Norway's Highest Ice Patches." *Royal Society Open Science* 5, no. 1: 171738.

Pilø, Lars, Thomas Reitmaier, Andrea Fischer, James H. Barrett, and Atle Nesje. 2022. "Ötzi, 30 Years On: A Reappraisal of the Depositional and Post-depositional History of the Find." *The Holocene* 33, no. 4: 095968362211261.

Putzer, Andreas, Daniela Festi, and Klaus Oeggl. 2016. "Was the Iceman Really a Herdsman? The Development of a Prehistoric Pastoral Economy in the Schnals Valley." *Antiquity* 90: 319–336.

Randall, Robert. 1982. "Qoyllur Rit'i, An Inca Fiesta of the Pleiades: Reflections on Time & Space in the Andean World." *Bulletin de l'Institut Français d'Études Andines* 11, no. 1–2: 37–81.

Reckin, Rachel. 2013. "Ice Patch Archaeology in Global Perspective: Archaeological Discoveries from Alpine Ice Patches Worldwide and Their Relationship with Paleoclimates." *Journal of World Prehistory* 26: 323–385.

Rodríguez, Jesús, Christian Willmes, and Ana Mateos. 2021. "Shivering in the Pleistocene: Human Adaptations to Cold Exposure in Western Europe from MIS 14 to MIS 11." *Journal of Human Evolution* 153: 102966.

Rosvold, Jørgen. 2018. "Faunal Finds from Alpine Ice—Natural or Archaeological Depositions?" *Journal of Glacial Archaeology* 3, no. 1: 79–108.

Rowlinson, J. S. 1998. "'Our Common Room in Geneva' and the Early Exploration of the Alps of Savoy." *Notes and Records of the Royal Society of London* 52, no. 2: 221–235.

Shaheen, Farhet A. 2016. "The Art of Glacier Grafting: Innovative Water Harvesting Techniques in Ladakh." Water Policy Research Highlight, International Water Management Institute.

Shelley, Mary. *Frankenstein: Annotated for Scientists, Engineers, and Creators of All Kinds*. Edited by David H. Guston, Ed Finn, and Jason Scott Robert. 2017. Cambridge, MA: The MIT Press.

Slawinska, Joanna, and Alan Robock. 2018. "Impact of Volcanic Eruptions on Decadal to Centennial Fluctuations of Arctic Sea Ice Extent during the Last Millennium and on Initiation of the Little Ice Age." *Journal of Climate* 31: 2145–2167.

Smith, Kristin. 2021. "Mass Balance, Accumulation Dynamics and High-altitude Warfare: The Siachen Glacier as a Battlefield." *Small Wars and Insurgencies* 32, no. 8: 1193–1220.

Smithsonian National Museum of Natural History. "Climate and Human Evolution." Updated May 4, 2022. https://humanorigins.si.edu/research/climate-and-human-evolution.

South Tyrol Museum of Archaeology. n.d. "Otzi the Iceman." https://www.iceman.it/en/the-iceman.

Spindler, Konrad. 1996. *The Man in the Ice: The Discovery of a 5,000-Year-Old Body Reveals the Secrets of the Stone Age.* 1st Canadian ed. New York: Three Rivers Press.

Spindler, Konrad. 1996. "Iceman's Last Weeks." *In Human Mummies: A Global Survey of their Status and the Techniques of Conservation,* edited by Konrad Spindler, Harald Wilfing, Elisabeth Rastbichler-Zissernig, Dieter zur Nedden, and Hans Nothdurfter: 249–263. Vienna, Austria: Springer-Verlag.

Stiebing, William H. Jr. 1993. *Uncovering the Past: A History of Archaeology.* New York: Oxford University Press.

Sun, Zixiang and Gang Qiao. 2021. "A Review of Surge-Type Glaciers." The International Archives of the Photogrammetry, Remote Sensing and Spatial Information Sciences XLIII-B3-2021: 503–508.

Swanton, John Reed. 1909. "Tlingit Myths and Texts." *Bureau of American Ethnology Bulletin* 39: 1–451.

Taillant, Jorge Daniel. 2021. *Meltdown: The Earth Without Glaciers.* New York: Oxford University Press.

Taylor, Rebecca S., Micheline Manseau, Cornelya F. C. Klütsch, Jean L. Polfus, Audrey Steedman, Dave Hervieux, Allicia Kelly, Nicholas C. Larter, Mary Gamberg, Helen Schwantje, and Paul J. Wilson. 2021. "Population Dynamics of Caribou Shaped by Glacial Cycles before the Last Glacial Maximum." *Molecular Ecology* 30: 6121–6143.

Taylor, William, Julia K. Clark, Björn Reichhardt, Gregory W. L. Hodgins, Jamsranjav Bayarsaikhan, Oyundelger Batchuluun, Jocelyn Whitworth, Myagmar Nansalmaa, Craig M. Lee, and E. James Dixon. 2019. "Investigating Reindeer Pastoralism and Exploitation of High Mountain Zones in Northern Mongolia through Ice Patch Archaeology." *PLOS ONE* 14, no. 11: e0224741.

Tolland, Andrew. 2015. "The Exorcism of Glaciers." *Scapegoat: Architecture, Landscape, Political Economy* 8: 30–45.

Tveiten, Ingvar. 2007. "Glacier Growing: A Local Response to Water Scarcity in Baltistan and Gilgit, Pakistan." Master's Thesis, Norwegian University of Life Sciences.

Vistad, Odd Inge, Line C. Wold, Karoline Daugstad, and Jan Vidar Haukeland. 2015. "Mimisbrunnr Climate Park: A Network for Heritage Learning, Tourism Development, and Climate Consciousness." *Journal of Heritage Tourism* 11, no. 1: 43–57.

Wang, Ke, Kay Prüfer, Ben Krause-Kyora, Ainash Childebayeva, Verena J. Schuenemann, Valentina Coia, Frank Maixner, Albert Zink, Stephan Schiffels, and Johannes Krause. 2023. "High-Coverage Genome of the Tyrolean Iceman Reveals Unusually High Anatolian Farmer Ancestry." *Cell Genomics* 3, no. 9: 100377.

World Glacier Monitoring Service. Updated February 2, 2023. https://wgms.ch.

Xu, Chi, Timothy A. Kohler, Timothy M. Lenton, Jens-Christian Svenning, and Marten Scheffer. 2020. "Future of the Human Climate Niche." *Proceedings of the National Academy of Sciences* 117, no. 21: 11350–11355.

Zink, Albert R. and Frank Maixner. 2019. "The Current Situation of the Tyrolean Iceman." *Gerontology* 65, no. 6: 699–706.

Index

INDEX

INDEX